The Executive Guide to Blockchain

"Blockchain and cryptocurrency are still mysteries to so many, even those in authority whose organizations depend on their knowledge and wisdom to guide their organizations into the future. *The Executive Guide to Blockchain* is an easy-to-read yet power-packed guide to increasing our knowledge of this technology, its impact on our global world and very importantly, its impact on the businesses and organizations we run. It's a timely must-read for anyone, especially those still yet confused by the technologies whose impact will be more than merely financial, but also monumental to our social fabric."

—Brenda Darden Wilkerson, *President and CEO of AnitaB.org, USA*

"There has been so much hype around blockchain and cryptocurrencies that this excellent book is a very timely guide to help the reader navigate through the complexities and make a proper assessment of whether the technology is relevant to them. The authors have a blend of business and technical backgrounds that enable them to communicate the key ideas in a readily understandable way. This book will be an essential read for busy executives who want to understand how blockchain can help them."

—Professor Christopher Hankin, *Former Co-director of the Institute for Security Science and Technology at Imperial College London, UK*

"I've really enjoyed reading *The Executive Guide to Blockchain* and hope you will too. Blockchain, cryptocurrencies and Bitcoin are not going away any time soon, understanding what they are and their effect on the world from a financial, business and social perspective is a must for anyone wanting to understand what is happening in the world today."

—Sue Black, OBE, *Professor of Computer Science and Technology Evangelist, Durham University, UK*

"A welcome, easy to follow guide that demystifies blockchain without assuming any previous knowledge of blockchain technology. I thoroughly enjoyed reading the authors' insights on why and how blockchain continues to be the great industry disruptor. *The Executive Guide of Blockchain* succeeds in covering all aspects of the application of blockchain in a compact yet clear way. An impressive book that will be a priceless tool to every business influencer. I highly recommend it."

—Ioannis Xanthakos, *Managing Director, Head of EMEA Equities Technology, Bank of America, UK*

"As two modern day Virgils, Maria Grazia Vigliotti and Haydn Jones guide us though the arcane of blockchain technology and cryptocurrencies. Explaining complex concepts in simple terms is not easy: the authors have accomplished this task brilliantly. *The Executive Guide to Blockchain* is not only a useful tools for business, but for the public at large, thus allowing ordinary people to harness the potentialities of new technologies."

—Massimo Carnelos, *Head of Economic Office at the Italian Embassy in London, Alternate Executive Director for Italy at the European Bank for Reconstruction and Development*

Maria Grazia Vigliotti ·
Haydn Jones

The Executive Guide to Blockchain

Using Smart Contracts and Digital
Currencies in your Business

Maria Grazia Vigliotti
Sandblocks Consulting
London, UK

Haydn Jones
Blockchain Hub
London, UK

ISBN 978-3-030-21106-6   ISBN 978-3-030-21107-3   (eBook)
https://doi.org/10.1007/978-3-030-21107-3

Cover credit: Andril Kolomiiets
Cover design by Fatima Jamadar

This Palgrave Macmillan imprint is published by the registered company Springer Nature Switzerland AG
The registered company address is: Gewerbestrasse 11, 6330 Cham, Switzerland

To Steffen, my love, my husband, my equal partner
—Maria Grazia Vigliotti

"Dedicated to my father—Derek Peter Jones, 1939–2019—a truly wonderful
man, who taught me to always do my best"
—Haydn Jones

Foreword by Sue Black

I can't remember when I first heard the terms cryptocurrency, bitcoin and blockchain, but it was quite some time ago. Maybe it was after hearing about the guy in California who bought 2 pizzas for ten thousand bitcoin, in May 2010. Who knows what he was thinking at the time, he obviously had no idea that bitcoin would really take off and increase in value to the extent that it has. Those two pizzas are probably the most expensive pizzas ever.

When Satoshi Nakamoto created the first-ever 50 bitcoins on January 3, 2009, he did it with the aim of creating a payment method run by people rather than faceless organisations. I say he but we even ten years later, we don't actually know who Satoshi Nakamoto is. Lots of effort have gone into finding out who Satoshi is, but as yet, no one knows. Well of course Satoshi knows, and maybe a few other people in Satoshi's confidence, but in general we don't know who Satoshi is, Satoshi is a real live, modern-day enigma.

There has been so much written about blockchain over the last few years, but nothing that I have seen that gives such a good, interesting and easy to read the explanation of blockchain theory along with understandable, implementable, practical advice. *The Executive Guide to Blockchain* also gives us a more global perspective of blockchain with real-world examples. Having travelled quite a bit over the last few years, its been really interesting to see how different making a payment can be from one country or even one region to another. Living in London, UK, I'm used to paying for practically everything on my phone using Apple Pay for purchases, and occasionally a credit card. I don't usually carry cash in London as there's no point, I don't ever need it. Working as a Professor at Durham University in the north of England for the last few months, I now need to carry cash with me as if I need to get a taxi in Durham I have to pay in cash. Travelling to Nigeria a couple of years ago for my daughter's wedding I found that it's not a good idea to use credit cards as they are not considered a secure method of payment there, all transactions were either paid in cash or by bank transfer.

Blockchain is not just about big business and investment opportunity. This book also tells us about how mobile banking is being used in Kenya via M-pesa which enables people without a traditional bricks-and-mortar bank account to be able to transfer even very small amounts of money for goods and services via mobile phones. M-Pesa is now being used in several other countries including Tanzania, Afghanistan, India and Romania empowering people on low incomes to easily transact with each other and the goods and service providers that they need to interact with without having to pay extortionate transaction fees or the inconvenience and insecurity of having to pay everything in cash.

It's great to read about how blockchain is being used by the World Food Programme (WFP) to handle payments for Syrian refugees in Jordan with hundreds of thousands of refugees receiving payments in this way. This has enabled a reduction in transaction costs of up to 98% and provides the WFP with an in-house record of every transaction.

There's also interesting information in this book about the future of blockchain. It's interesting to hear that Facebook is planning on launching a cryptocurrency in 2020 which will enable 1.7 billion people worldwide who don't have bank accounts to exchange money. This will be a great step forward in empowering people on low incomes around the world, helping them to help themselves. There's another discussion to be had around whether Facebook is the best platform to implement this service.

Overall this book is a clear and concise guide which makes good use of lots of diagrams and examples to help explain everything. It has a real practical focus, for example Six-step Blockchain Strategy, while also explaining the theory well. Like lots of things, blockchain is very simple and very complicated at the same time.

There's also smart, but not always followed, high-level technology-related advice like remember that technology supports the business, not the other way around. So many times throughout my career I've seen companies wanting to implement cool new technology because they feel like everyone else is doing it and they don't want to get left behind. However, used to its best advantage technology provides us with a suite of tools to solve business problems. The business strategy should come first with the technology there to support the goals of the organisation, and as such should be completely aligned with the business strategy.

"I've really enjoyed reading *The Executive Guide to Blockchain* and hope you will too. Blockchain, cryptocurrencies and Bitcoin are not going away any time soon, understanding what they are and their effect on the world from a financial, business and social perspective is a must for anyone wanting to understand what is happening in the world today."

Durham, UK   Sue Black
July 2019   OBE, Professor of Computer Science
                                                     and Technology Evangelist
                                                           Durham University

Foreword by Professor William Knottenbelt

The brave new world of cryptocurrencies, blockchains, distributed ledgers and smart contracts is well and truly upon us, with new decentralised technologies and applications being launched daily around the globe. Yet the breakneck pace of technological advance and a growing torrent of specialist terminology can make this frontier of technology seem impenetrable for all but the most dedicated technophiles and scholars. This is where *The Executive Guide to Blockchain* comes in. In their inimitable style that is simultaneously lively and elucidating, the authors cut through the flood of baffling jargon to distil succinctly not only the technical essence underpinning blockchain technology, but also the potential for it to transform business models and practices. Importantly, the book is balanced in its advocacy, giving careful consideration to when it might be—and when it might not be—appropriate to use blockchain technology for particular use cases, often highlighting real-world examples. It also sheds light on vital non-technical considerations such as legal and regulatory aspects.

I hope you will find this timely book to be as insightful and illuminating as I did and that it will equip you with a new understanding and working knowledge that you can put into practice where it really matters: in the real world.

London, UK
July 2019
Professor William Knottenbelt
Director Centre for Cryptocurrency
Research and Engineering, Imperial College London

Preface

A new technology has landed and it is having a significant impact on the way business is conducted across the world. It has a number of guises: you may know it as blockchain; or you may be more familiar with the emergence of digital money, aka cryptocurrency. It also goes by the more prosaic title of Distributed Ledger Technology (DLT). But, regardless of the terminology used to describe this technological breakthrough, what you will undoubtedly want to know is: What does this mean for me and my business? The potential is huge, and it is creating ripples throughout the business world. If you haven't already heard how your business could benefit from *blockchain*, it's likely you soon will. Of course, you might already know someone who has left a well-paid job to run his or her own blockchain or cryptocurrency start-up, or heard of people trading in cryptocurrency who are making returns that experienced fund managers simply can't explain. The Economist, Financial Times and other serious newspapers have all published articles on blockchain and cryptocurrencies. Government Departments and regulators around the world, have issued reports or guidance about this rapidly emerging technology. There are numerous stories about new crypto-millionaires although some financial industry figures have urged caution, warning of a potential bubble or, more worryingly, a 'Ponzi scheme'.

This book provides a measured, nuanced view of blockchain and its associated technologies by clarifying and systematising a broad range of topics. It sets out to:

1. Explain for readers with little or no computing knowledge how blockchain technology works
2. Show the benefits of this technology beyond the financial sector

3. Discuss the benefits and the risk of smart contracts
4. Examine current regulation
5. Provide the tools to develop a comprehensive business strategy.

It also looks at the potential for blockchain technology in the next decade, based on the authors' experience in the sector, and it collects in one place a significant amount of resources to understand how blockchain technology is deployed around the world.

Each chapter takes a specific topic and attempts to answer frequently asked questions. A glossary at the end of the book is a useful tool to help the reader navigate this exciting new world.

As the blockchain sector matures, and new regulation and business models emerge, some topics in this book will become obsolete. To keep you updated about the very latest developments, we have created an associated website https://www.sandblocksconsulting.co.uk/.

Why Us?

Maria and Haydn have a unique set of business and technical skills. We have experience in banking, software development, cybersecurity and cryptography, digital innovation and payment systems, and we have led blockchain projects across a range of sectors. In the last few years, we have assisted businesses worldwide by:

- Developing blockchain strategies
- Delivering blockchain projects
- Advising on the benefits and risk of the technology
- Training employees in a wide range of sectors
- Writing secure by design *smart contracts*
- Auditing smart contracts for security purposes.

By sharing our experience, we aim to:

- Make the technology accessible for a non-technical business audience
- Share best practices in developing blockchain projects
- Combine useful information from different sources.

We hope you enjoy this book, and that it gives you the confidence to use blockchain to improve your organisation's technology and help reduce operational costs.

London, UK   Maria Grazia Vigliotti
July 2019   Haydn Jones

Acknowledgements

This book was completed with the help of several people, who, within their busy life and out of the goodness of their heart, have dedicated time and resources to read and comment early drafts.

I like to thank my family: Steffen, Martha and Isaac. Steffen was there when emotional support was needed; he deserves all the credit for having typeset the internal look of the book to make it more appealing, and provided a plenty of invaluable feedback on the use of English and on the content. Martha was one of the first enthusiastic readers, and her astute questions help me to think about the book from different prospectives. Isaac patiently listened the ideas planned for the book and cheered me up at difficult times.

To 'the early readers', the friends and colleagues, who provided detailed and professional feedback: I like to thank you for your support.

Sarah Hewin
Steve Webb
Luca Viganó
David G. W. Birch
Daniel Mermelstein
Alberto De Capitani
Chris Wray
Jonathan Price
Valerie Kahn
Debbie Tanner
Natalie Hankey

Eva Kingston
Daren Foresyth
Akber Datoo
Roman Gonitel

I'm deeply indebted to Gabriel Everington at Palgrave for his editorial help, ongoing support and boundless patience.

Maria Grazia Vigliotti

Contents

List of Figures

1

Introduction

Most of us know people who resort to using jargon in an attempt to make themselves look more knowledgeable or important. It's hugely irritating and the inevitable result is confusion and an abject failure to explain. The renowned American computer scientist and *Turing Award*[1] winner Leslie Lamport[2] shunned jargon and on some occasions just used clocks to explain highly complex computing problems. Throughout this book, we have adopted a similar uncomplicated approach to show how blockchain technology and cryptocurrencies are changing the way we work.

Whether you are a highly experienced executive or a recent graduate (or anything in-between), this book, with its chapter-by-chapter demystification of blockchain and cryptocurrencies, will have something of interest and improve your knowledge of the technology. Differently from other books, we do not merely focus on the business' benefits and leave details of technology to the 'experts', nor we focus only on the technology, leaving unclear how and when the benefits can be delivered, but aim to give balanced attention to both.

You may, of course, already be familiar with blockchain or Bitcoin. If most of what you know has come via the Internet, don't worry, you're in good company! Google Trends[3] revealed that in 2018 'What is Bitcoin?' was the

---

[1]The A.M. Turing Award, sometimes referred to as the 'Nobel Prize of Computing', was named in honour of Alan Mathison Turing (1912–1954), a British mathematician and computer scientist. He made fundamental advances in computer architecture, algorithms, formalisation of computing and artificial intelligence. Turing was also instrumental in British code-breaking work during Second World War (source https://amturing.acm.org).

[2]For more information about Lamport's work, see https://amturing.acm.org/award_winners/lamport_1205376.cfm.

[3]Google Trends is an open site owned by Google that publicises searches made by users from 2004, https://trends.google.com/trends.

© The Author(s) 2020
M. G. Vigliotti and H. Jones, *The Executive Guide to Blockchain*,
https://doi.org/10.1007/978-3-030-21107-3_1

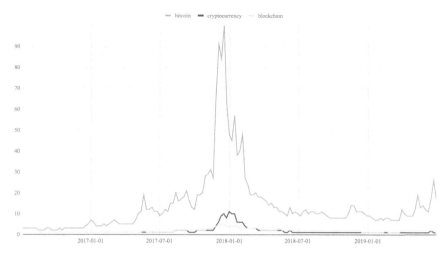

Fig. 1.1 Comparative display of amount of Google searches carried out worldwide on the words 'Bitcoin', 'Cryptocurrency' and 'Blockchain' from July 1, 2016 until June 30, 2019 (*Data source* Google Trends, https://trends.google.com/trends)

most asked question on Google[4] in the USA that year, suggesting a great interest in the subject. In the EU, 66% of people have heard of cryptocurrency, even though only 9% of them actually own a cryptocurrency [89].[5] Ownership of a smartphone or tablet is driving knowledge of cryptocurrencies with older people showing as much interest as those between the ages of 25 and 40. In the future, one in four expects to own a cryptocurrency, and about a third believes that Bitcoin is the future of spending online [89].

You may be curious about the worldwide popularity of cryptocurrencies, Bitcoin or blockchain. Google Trends is useful to identify the geographical interest in the technology, and sufficiently reliable in its results: it has been used for scientific research across various fields [82]. When Bitcoin's value peaked in December 2017, there was a corresponding spike in Internet searches, which is probably not surprising (see Fig. 1.1). To a lesser extent, people seem to be interested in understanding what are cryptocurrencies and blockchain as well. The strongest interest in Bitcoin came from Nigeria, South Africa and Ghana (see Fig. 1.2), while the highest number of searches for blockchain originated in China (see Fig. 1.3). This is not surprising given that China is leading the world in the use and development of blockchain technology and has filed the most patents related to blockchain in the world. The most searches

---

[4]See the voice 'What is Bitcoin' at https://trends.google.com/trends/yis/2018/US (accessed February 20, 2019).

[5]In the study, cryptocurrency is defined as 'a kind of digital currency not created or secured by the government but by a network of individuals'.

Fig. 1.2 Comparative display of amount of searches carried out on word 'Bitcoin' from July 1, 2016 until June 30, 2019 displayed as percentage of the highest (*Data source* Google Trends, https://trends.google.com/trends)

Fig. 1.3 Comparative display of amount of searches carried out on word 'Blockchain' from July 1, 2016 until June 30, 2019 displayed as percentage of the highest (*Data source* Google Trends, https://trends.google.com/trends)

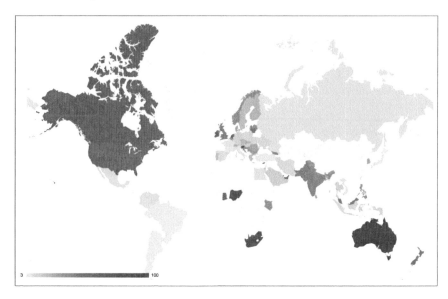

Fig. 1.4 Comparative display of amount of searches carried out on word to 'Cryptocurrency' from July 1, 2016 until June 30, 2019 displayed as percentage of the highest (*Data source* Google Trends, https://trends.google.com/trends)

for cryptocurrencies came from Slovenia (see Fig. 1.4). These results indicate interest in different aspects and application of the same technology.

A Driver for Change—Or a Bubble About to Burst?

Since the first cryptocurrency, Bitcoin, was created a decade ago, opinion has been polarised about its benefits. Some investors, including a 12-year-old boy, have become millionaires.[6] A $1000 investment in bitcoin at the beginning of 2013 would have been worth more than $471,000 in April 2019, an incredible return by any measure. The media have compared the volatility of cryptocurrencies to the infamous Dutch seventeenth-century 'Tulip Mania', generally considered to be the first-ever speculative bubble [81]. The pessimists fear the sudden drop of prices (the market fell approximately 90% between the end of December 2017 to December 2018) are the precursors of the next big financial crisis. There is, however, no evidence to back this theory. Even at its peak in December 2017, when the cryptocurrency market was estimated to be

---

[6]See the article on the *Guardian* in June 2018, https://www.theguardian.com/technology/2018/jun/13/meet-erik-finman-the-teenage-bitcoin-millionaire (accessed April 10, 2019).

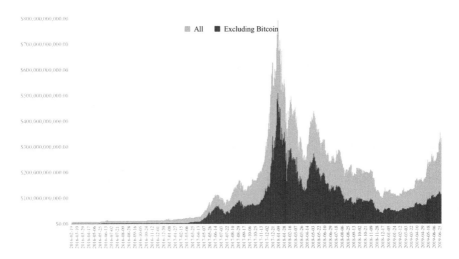

Fig. 1.5 Graph of historical value of the market capitalisation of cryptocurrencies. At the highest peak towards on December 21, 2017, it was estimated to be worth approximately $800 billion (*Data source* Coin Dance, https://coin.dance/stats/marketcaphistorical)

approximately $800 billion, it was less than 1% of global *GDP* (see Fig. 1.5). By contrast, at the peak of the dot-com bubble in 2000, the combined market capitalisation of US technology stocks was close to a third of world GDP; and in 2008, with the global economy teetering on the edge of financial meltdown, the notional value of credit default swaps was a staggering 100% of world GDP [43]. Nevertheless, it's a fact that some very influential people do have different views about cryptocurrencies. John McAfee, the security expert and creator of the first commercial antivirus software, is a vocal supporter, while Warren Buffet, the American business magnate and investor, suggests to 'stay away from it'.

Giving You the Facts—Chapter by Chapter

With such a range of views, it's important that readers are able to form their own opinions and consider the impact both blockchain and cryptocurrencies could have on business.

We have set out to provide clarity in five areas:

History Looking at the origin and development of cryptocurrencies
Technical A jargon-free explanation of how blockchain technology and smart contracts work

Economic How cryptocurrencies can be used to buy goods and services

Business Building a digital strategy that includes blockchain and cryptocurrencies and an understanding of relevant case studies

Legal Look at how cryptocurrencies are regulated and future regulation of smart contracts.

The history of cryptocurrencies, covered in Chapter 2, goes beyond the hype and examines the facts, with an explanation of the development of both blockchain and cryptocurrencies.

Chapter 3 examines cryptography, a word to strike horror in the hearts of non-mathematicians but which we think you will come to enjoy.

Understanding how Bitcoin and blockchain work is made simple in Chapter 4, where we also examine the history of Bitcoin and the reasons for the attention since it was created in 2009.

Chapter 5 offers an overview of the cryptocurrencies on the market and practical advice on how to use them to buy goods and services. It also looks at international payments using cryptocurrencies in countries with limited banking infrastructure. It also examines the various kinds of cryptocurrencies (stable coins, private coins, tokens, etc.) and provides the reader with a road map to understand this complex landscape.

The evolution of blockchains from open networks run by communities to enterprises operated by federations or large organisations is looked at in Chapter 6. Application of blockchain solutions for enterprises is thoroughly explored.

For all its benefits, blockchain isn't suitable for everyone. Chapter 7 provides practical advice on how to implement a blockchain strategy for businesses of any size.

The history and current use of smart contracts is put under the microscope in Chapter 8. If you work in computer science or the legal sector, you won't want to skip this chapter.

As blockchain enters its second decade, there are calls for greater regulation. Chapter 9 considers how over-regulation could stifle innovation, while too little could have damaging consequences.

Finally, in Chapter 10, we dust off our crystal ball and look at the future for both blockchain and cryptocurrencies, including whether central banks will get on the board of the cryptocurrency train.

We hope you enjoy reading this book and gain a greater understanding of the technology underpinning blockchain and cryptocurrencies, as well as the significant impact both could have on global business. Good luck and good reading.

2

Beyond the Hype

On January 3, 2009 the cryptocurrency era began, without, it has to be said, fanfare or razzmatazz. Nevertheless, on that day 50 bitcoins were privately created in the first block of the first ever *blockchain*, or what today is known as the *genesis block*.

What marked out *Bitcoin* from the start—what made it different, in fact—was that no bank, government, payments-system provider, global technology company or regulator was involved in its incarnation. Indeed, its creator, who used the pseudonym *Satoshi Nakamoto*, was determined to do away with the need for any kind of payment-enabling third-party or financial institution.

Nakamoto marked the creation of the ground-breaking cryptocurrency with an encoded message: '*The Times 03/Jan/2009 Chancellor on brink of second bailout for banks*'. By encoding a headline from a national newspaper, Nakamoto had cleverly devised a simple and efficient *time-stamping* procedure,[1] which showed the block could not have existed before the publication date of the newspaper article.

A decade later, Bitcoin has grown to become easily the most recognised of the thousands of cryptocurrencies in existence. Crucially, it still doesn't depend on intermediaries, such as banks, to process payments; unlike conventional currencies, its supply isn't controlled by a central bank. What it does have, however, is a market valuation because bitcoins (and some of the other cryptocurrencies) can be used to buy goods and services.

---

[1]Using a newspaper as time-stamping procedure is not new. In certain kidnapping cases, to provide evidence that victims were alive at a specific point in time, a picture of them with the a newspaper was taken and sent to families.

© The Author(s) 2020
M. G. Vigliotti and H. Jones, *The Executive Guide to Blockchain*,
https://doi.org/10.1007/978-3-030-21107-3_2

Bitcoin: A Currency Born Out of Crisis

Bitcoin was created in the wake of the 2008 global financial crisis, arguably the biggest shock to the world economy since the great depression of the 1930s. Inevitably, the crash raised serious questions about the ability of governments and central banks to effectively regulate the financial sector; it also resulted in widespread mistrust of banks and bankers, who quickly became public enemy number one.

Nakamoto's original aim was for Bitcoin to be an efficient electronic payment method run by people rather than faceless organisations [94]. This chimed perfectly with anti-establishment anger that was prevalent in the immediate aftermath of the financial crash. Indeed, grass-roots activist movements around the world were happy to make Bitcoin the symbol of an entirely new financial system. *Cypherpunks*—activists, advocating use of strong cryptography and privacy, enhancing technologies to achieve social and political changes [86, 93]—even helped with the development of Bitcoin software in the early stages [9]. The link between the Cypherpunks and Bitcoin development reinforced the anti-establishment sentiment.

Hitting a Bump in the Road

The idealism that characterised the start of Bitcoin received an early setback when the fledgling currency was used for trading illegal drugs on the *Silk Road marketplace*, the first-known example of a hidden or *Darknet* market. Launched in February 2011, the site was eventually closed down in October 2013 by the FBI, which seized 144,336 bitcoins with an estimated value of $48 million [1]—an incident dramatised in an episode of the popular TV series The Good Wife.[2]

Although the site made a substantial amount of money, its founder, Ross Ulbricht paid a heavy price. The man who operated using the pseudonym 'Dread Pirate Roberts'[3] was sentenced to life in prison with no possibility of parole. Since then there have been further incidents where Bitcoin and other cryptocurrencies have been used to finance crime [80] or for money laundering [40]. But while these instances can't and shouldn't be ignored, it's important to

---

[2]Episode 59 first aired on January 15, 2012.

[3]Named after a pirate of near-mythical reputation, a fictional character from William Goldman's novel *The Princess Bride* (1973) and its 1987 film adaptation.

put them in perspective. Established financial institutions often fail to prevent illegal activity such as money laundering.[4]

Taking a balanced view of cryptocurrencies is probably the best way forward for potential adopters. The hugely optimistic claims made by some digital currency evangelists should be viewed with a healthy dose of scepticism. But the potential for good undoubtedly exists. It's worth recalling that in its early years, the World Wide Web attracted some dubious users; today most of us would admit that we probably couldn't function without it. We may never feel quite like that about cryptocurrencies but as they mature, and their use grows, regulators are set to play an increasingly important role in reducing illicit use, as we will see later in this chapter.

Expansion

In the ten years since its creation, bitcoin has gone from being a hobby for a small group of enthusiastic cryptographers to the inspiration behind the creation of thousand more cryptocurrencies with an estimated capitalisation at the of June 2019 of more than $343 billion.[5] That's the equivalent of the GDP of Denmark! Figure 2.1 provides a useful comparison of the cryptocurrency market against the total amount of money in circulation.

There are more than two thousand online cryptocurrencies,[6] and over seven hundred so-called crypto-funds (i.e. hedge funds that mainly invest in cryptocurrencies or blockchain-related companies) have emerged, attracting well over $14 billion in assets under management [7].

It is predicted that in the period 2018–2022, companies will spend a remarkable $12.4 billion in blockchain [58]; Fig. 2.2 shows industry spending in 2018. If this trend continues, blockchain's impact will substantially extend beyond the financial sector.

---

[4]See   https://www.theguardian.com/business/2019/apr/17/deutsche-bank-faces-action-over-20bn-russian-money-laundering-scheme (accessed April 17, 2019).

[5]At the time of writing on June 30, 2019.

[6]See for example https://coinmarketcap.com/currencies/volume/24-hour (accessed June 30, 2019).

Market Capitalisation Comparison

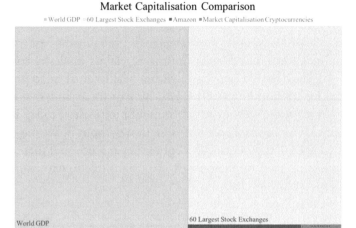

Fig. 2.1 Comparison among the market capitalisation of cryptocurrencies (June 30, 2019), Amazon (the world's biggest company), the amount of money in the 60 biggest stock exchanges, and world GDP, the total amount of money in circulation

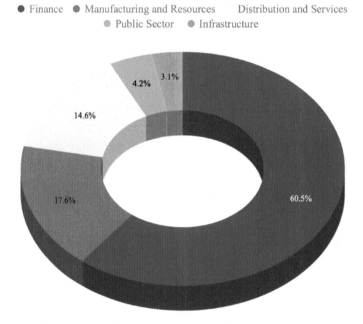

Fig. 2.2 Blockchain spending by industry in 2018 (*Source* IDC 'Word Wide Semi-annual Blockchain Spending Guide' 2017–2018-H1, https://www.idc.com)

Blockchain—A Success Story?

> **What Is a Blockchain?**
>
> A blockchain is a system of authoritative electronic records shared by independent entities via an automatic consensus process, dispensing of a central co-ordinator party.

The growth of blockchain technology has matched and arguably exceeded the rapid expansion of cryptocurrencies. In the Middle East, blockchain is growing exponentially, with Dubai promising to put all government services on the blockchain by 2020 [68]. Europe, too, has seen rising enthusiasm for blockchain, although some countries are taking a more measured approach towards its development.

The EU has developed two initiatives: the European Blockchain Partnership, which enables the 27 countries to collaborate on regulatory and technical matters, and the European Blockchain Observatory, with an investment of €300 million over three years to undertake research on how blockchain can be applied. In the UK, for example, the *financial regulator*, the *Financial Conduct Authority* (FCA), has created a *Regulatory sandbox* [20], a safe place where business, working with innovative financial products that involve blockchain, can test their product with real consumers under the watchful eye of the regulator. In this way, the regulator has first-hand experience of the benefits and risks of the new technology.

Around the world there are many more examples of the impact blockchain is having on business. The *Monetary Authority of Singapore* (MAS) and the country's stock exchange have launched a prototype method for delivery, payment and settlement of assets, using blockchain [17]. Global financial institutions like J.P. Morgan [13] and Banco Bilbao Vizcaya Argentaria (BBVA) [4, 5] are investing in the technology while Santander [21] and HSBC [11] are deploying the technology to improve cross-border payments.

Software giants such as IBM, SAP, Microsoft and Oracle all offer blockchain solutions to their clients. Facebook is shaking up the banking world with the plan to issue its own cyrptocurrency, Libra [18]. Given this growth, one would assume that both blockchain and cryptocurrencies are here to stay. To assess whether or not that is the case, we need to take an in-depth look at how they work and the benefits they deliver.

What Are Cryptocurrencies?

Once we had travellers' cheques, now we have cryptocurrencies. That might be an oversimplification, but there are certainly similarities in the way both methods of payments work. A travellers' cheque allows payments from one person to another across currencies. Usually issued by banks, their unique selling point (USP) is that they can never bounce because they've already been paid for. In terms of how they are used, each cheque has a legitimate sequence of signatures; the first belongs to the issuing bank, which in effect is a promise to pay the value to the rightful owner. The similarities between travellers' cheques and cryptocurrencies are:

- Both have a sequence of signatures.
- Both are a form of payment, with built-in protection against fraud.

But while a bank acts as the guarantor of a travellers' cheque, cryptocurrencies are run by an *algorithm* on a *network* of computers: there's no single organisation enforcing correct behaviour; they are, in the jargon, *decentralised*. A set of rules inserted in software determines how cryptocoins are minted and transferred and, typically, how many can ever exist (this is usually a fixed number). Digital signatures showing change of ownership of cryptocurrencies are underpinned by cryptography. From this, it's possible to see four main differences between traveller's cheques and cryptocurrencies:

1. Travellers' cheques are issued and used in paper format; cryptocurrencies are entirely digital, including the signatures.
2. Travellers' cheques are issued by a bank; cryptocurrencies are maintained by a community.
3. Travellers' cheques are trusted because they are backed by banks; cryptocurrencies are accepted because people believe in the technology and, in some cases, in *decentralised* ideology.
4. The signature on a travellers' cheque reveals the owner's identity; cryptocurrencies are digital signatures do not reveal the owner's identity.

Are Cryptocurrencies Really Money?

We all use *money*. But how can we define what it is? Of course, we know it's the way we pay for goods and services. But it's also a *unit of account* that enables us to compare, for example, one hour's work provided by a sought-after consultant with a meal in a Michelin star restaurant. It is also a *Medium of exchange*: the consultant passes some or all of the money he receives for his services to settle his bill in the restaurant. Finally, it is a *store of value*; the restaurant uses the money it receives as payment for meals to purchase other goods or services. In modern economies, currencies issued by central banks, all have these three functions (Fig. 2.3). However, these functions can also be met without money being exchanged. In prisoner of war camps in the Second World War, cigarettes fulfilled all the three functions; however, nobody would compare cigarettes to currency [59].

Can cryptocurrencies be regarded as money in the conventional sense, meeting the functions outlined above? The Bank of England doesn't think so [43, 66]. It says cryptocurrencies are:

1. Too volatile to be a reliable store of value, see Fig. 2.4 which shows that Bitcoin can reach a daily change of over 20%. In comparison, a major event as Brexit, which shook the financial market, led to a change of 7.2% in the exchange of US dollar against the British Pound.
2. An inefficient medium of exchange which is not yet widely accepted,[7] although prominent retailers including Microsoft, Subway and Virgin Galactic accept payment in Bitcoin and Starbucks is set to follow.

Fig. 2.3 The three functions of money—from the most to the less important

---

[7]The website SpendaBit http://spendbitcoins.com claims that more than 100,000 merchants accept Bitcoin.

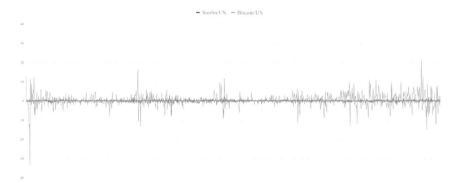

Fig. 2.4 Comparison of daily change of price of Bitcoin in US dollar, and British Pound in US dollar from December 1, 2014, until May 11, 2019 (*Source* https://data.bitcoinity.org/markets/price/5y/USD/coinbase?f=m5&r=day&t=l and Bank of England, https://www.bankofengland.co.uk/boeapps/database/default.asp)

3. It is not a unit of account because the first two functions are not well exercised.

Taking a wider view, there is a difference between *legal tender* and *common tender*. The former enables legal settlement of debts that can't be contested in a court. Examples are banknotes issued by the Bank of England and Royal Mint coins. Currently, no cryptocurrency is considered legal tender anywhere in the world. But businesses and people can accept other means of payment that are not legal tender, for example credit cards or foreign currency; or commodities like gold; virtual assets such as rewards points from reputable retailers; or cryptocurrencies, like bitcoin.

The way bitcoin acquires value is similar to a fiat money; quite simply, people trust it as a valid method of payment. Conventional currencies (in other words: money) are trusted by consumers because they are underpinned by central banks and their value is also dependent upon trust placed in the political and economic health of the country issuing the currency. Trust in cryptocurrencies comes mainly from people who understand how the technology works. To prevent inflation and preserve value, all cryptocurrencies restrict the number of coins that are in circulation.

There is a common misconception that money issued by central banks is backed by commodities, such as gold. In fact, this practice ended almost half-a-century ago, when the US Federal Reserve scrapped the *gold standard*, the agreement that currency could be converted to gold, in 1971 [67].[8]

---

[8]Gold standard refers to the Bretton Woods Agreement in 1944, when 44 nations agreed to adopt the US dollar as an official reserve currency and pegged the exchange rates for their currency to the US dollar.

Given their relatively short history, the rise in popularity of cryptocurrencies is remarkable. A recent study [84] shows that at least 139 million user accounts have been created with at least 35 million different users. However, of these only an estimated 11% actually use a digital currency to make a payment; the rest of the users deploy cryptocurrencies for speculation.

Are Cryptocurrencies Legal in the USA?

Surprisingly, it's not a crime to coin money in the USA. Article 1, Section 8 of the Constitution [6] states that counterfeiting can be prosecuted, and Section 10 prohibits Federal States coining money. However, private individuals and businesses can, so long as their money cannot be confused with state-issued currency. Bernard von NotHaus, the creator of the *Liberty Dollar*, was convicted in 2011 after printing and distributing his gold-backed currency. The *Liberty Dollar* was very similar in appearance to the US dollar, which leads to his conviction. As Bitcoin and other cryptocurrencies can't be confused with US currency, they are not banned under the US constitution.

Opinion is divided about the likelihood of *digital currencies* ever replacing conventional currencies or supplanting central banks. Instances of countries issuing cryptocurrencies are rare and, in the case of Venezuela, seemingly unsuccessful: in 2018, the government in Caracas launched the *Petro*, a national cryptocurrency backed by the country's oil reserves [19]. But despite claims that the Petro would be used for international transactions, there is little evidence of this happening and it has failed to combat Venezuela's rampant inflation [22].

A question of whether central banks should issue their own central bank digital currency is rather delicate, and the opinions on this matter are polarised: the UK does not see the need in the short term [43, 66], while the EU is investigating the benefits and risks [76]. Some central banks [52, 65], including the South African [46], the Brazilian [53] and Thailand's [3] are looking at piloting digital currency or variations of distributed ledger technology, indicating that the technology is influencing the financial heart of some countries. Despite the widespread use of cryptocurrencies around the world, some countries have either directly or indirectly banned their use (see Fig. 2.5). Others, including the USA, regard cryptocurrency as a commodity similar to gold, while the European Union categorises them as virtual money.

The agreement entailed that US currency could be converted to gold, which was cancelled by Nixon in 1971. The decoupling of the US dollar from the gold from the Federal Reserve happened in 1976, while the Bank of England decoupled the UK Pound Sterling from gold in 1973.

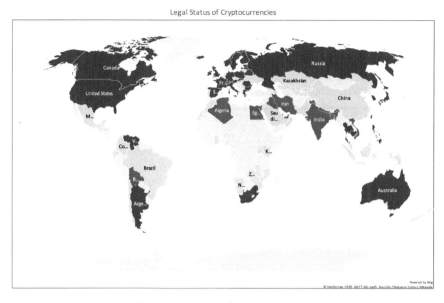

Legal Status of Cryptocurrencies

Fig. 2.5 Map describing the legal status of cryptocurrencies around the world. The countries coloured green formally or informally recognised cryptocurrencies as virtual currencies, or a form of payment or some taxable transfer of value; countries coloured red are directly or indirectly banned cryptocurrencies, and the countries coloured grey either there is not guidance or regulation, or the situation is unknown or ambiguous (*Source* The Law Library of Congress, https://www.loc.gov/law/help/cryptocurrency/world-survey.php)

Of course, it's entirely possible that cryptocurrencies will remain a niche market, although current trends point towards further growth and more widespread use. Key to their success is an efficient digital ecosystem that meets users' *needs*. Let's not forget, Bitcoin was originally created with the aim of improving electronic payments and there is still a compelling case for this today. The global *remittance* market is worth well over $600 billion annually [32, 79]; however, low-volume or small-value cross-border payments are poorly served by existing banking infrastructure and this is opening the way for start-ups using blockchain technology that is both more efficient and less expensive.

In the developed world, there has already been a significant decrease in the use of cash and a rise in digital payments. Fintech start-ups like Monzo,[9] Revolut[10] and Tide[11] are poaching customers from traditional banks with the lure of reduced charges and a new type of banking model that is attractive to young people. Mobile operators (not banks) enabling banking services without brick

---

[9]See https://monzo.com (accessed April 5, 2019).
[10]See https://www.revolut.com (accessed April 5, 2019).
[11]See https://www.tide.co (accessed April 5, 2019).

and mortar are making inroads: the blueprint is M-pesa,[12] which launched in Kenya in 2007, and has since spread to Tanzania, Afghanistan, South Africa, India, Romania and Albania. M-pesa allows users to deposit, withdraw, transfer money, borrow and pay for goods and services easily with a mobile device, yet without necessarily hold bank account, or ever having to see a clerk in a branch. It serves also a low-income population, living in remote and rural areas, in need of frequent, and (often) low-value transactions. And in China, Alibaba, the multinational conglomerate, dominates the mobile payments services.

Even if cryptocurrencies are not directly the solution to the problem, blockchain probably is. An example of how blockchain can help in cross-border payment comes from the World Food Programme (WFP) [51], which developed a blockchain-based Building Blocks system aimed at handling payments for food aid for Syrian refugees in Jordan. WFP assistance is increasingly being delivered in the form of cash transfers, which was around $1.6 billion in 2018. Blockchain acts as a trusted way to keep track of payments. In January 2017, WFP initiated a 'proof of concept' in Sindh province, Pakistan. It built and implemented a robust blockchain system in refugee camps in Jordan. By October 2018, more than 100,000 people residing in camps redeemed their WFP-provided assistance through the blockchain-based system. Thanks to the technology, WFP has a full, in-house record of every retail transaction, ensuring greater security and privacy for the Syrian refugees and a reduction of transaction costs of up to 98%. For refugees in camps, Building Blocks system has integrated with the United Nations High Commissioner for Refugees' (UNHCR) existing biometric authentication technology allowing them to provide instantaneous proof of identity. Building Blocks runs on a private blockchain controlled by WFP. The blockchain settlement system is more efficient than traditional solutions, is freely available and can be adapted to small and large cases.

Blockchain Gets Smart

The development of blockchain has led to an increase in the popularity of so-called *smart contract*s (see Chapter 8), which are valued for their transparency.

In the finance sector, smart contracts are being considered for financial derivatives, such as options and swaps. The efficiency they deliver helps to simplify the execution of complex financial transactions. The International Swaps and Derivative Association (ISDA) has worked on creating a framework

---

[12]See https://www.safaricom.co.ke/personal/m-pesa.

to standardise smart contracts [34, 47], and the likelihood is that a number of banks will follow this example.

What Is a Smart Contract?
A smart contract is a software application on the blockchain that automatically executes agreements among parties: for example, a smart contract can automatically trigger a payment when goods are delivered on time, or when the conditions for a future contract are met.

Away from the financial sector, blockchain is also having an impact on supply chains and trade finance [56]; management of digital rights in the creative sector [97]; donation and spending transparency in the charitable sector [51]; and authentication of academic certification.[13] In October 2015, the headline on the front page of the *Economist* newspaper read: 'The Trust Machine' [29] Inside, it examined the way blockchain is fostering trust between organisations who might otherwise view each other with suspicion; blockchain allows them to share data without ceding control to one party.

Throughout this book, we look at where this 'trust machine' could prove useful. One area is land registry in the developing world, where the unreliability of land registry is holding back the mortgage market and opportunities for landowners to borrow money against their property. The economist Eduardo De Soto [103] estimated that the value of unrecognised land in the world is about $9.3 trillion. In a number of developing countries, government land registry continues to hold paper records and land certificates are often forged with the collusion of poorly paid civil servants. Blockchain technology, however, enables participants, such as banks, to cooperate to keep a private land registry once ownership of land has been ascertained. The need is to ensure that the original data inserted in the blockchain land registry is correct; then, provided the consortium is large enough and the blockchain is accurately

---

[13] See for example https://gradba.se/en/.

implemented, the records can be trusted by any party querying the ledger. This revolutionary and cost-effective method of implementing blockchain software is available with very generous open-source license conditions. The challenge, however, is to bring the consortium together and to foster the collaboration.

Technology alone can't solve all business problems. For this reason, often entire sectors (banking, insurance, supply chain, etc.) support blockchain projects: this is the case for the banking industry with the R3 consortium[14] and the Corda blockchain and for the insurance sector with the B3i consortium.[15] As businesses learn to trust blockchain, the number of consortia using it is likely to grow. According to a recent survey [38], 29% of companies have already joined an existing consortium, and 45% claim to be likely to take part in one. This is recognition, perhaps, of the numerous benefits blockchain brings, including:

- Reduced costs for businesses,
- Facilitation of working with unified industry standard, and
- Fostering collaboration.

Regulation

With the rising use of cryptocurrencies and the widespread development of blockchain, the issue of regulation can't be ignored. Legislation already exists covering electronic and digital systems, such as the Computer Misuse Act 1990 [24] in the UK. But there is a strong argument for increasing regulation to combat potential fraud as blockchain technology grows and becomes more extensively used. Malta is one country that has moved to protect consumers and investors from false claims made about blockchain. In August 2018, it created the Malta Digital Innovation Authority [35], and the legislation[16] certifies blockchain applications. The new regulatory body supports the economic expansion of business while at the same time reducing risk. An exciting development in recent years has been the ability of companies to raise capital in exchange for new cryptocurrencies through a mechanism known as an Initial Coin Offerings (ICOs). ICOs, sometimes referred to as 'cryptocurrency crowd

---

[14] See https://www.r3.com (accessed May 10, 2019).
[15] See https://b3i.tech/home.html (accessed May 10, 2019).
[16] Schedule 2 [36] (accessed May 10, 2019).

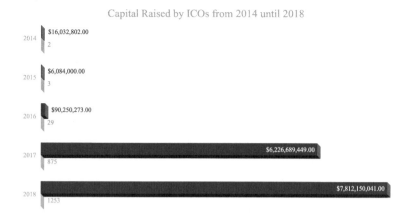

Fig. 2.6 Number of Initial Coin Offerings for each year (orange) and capital raised (blue) each year since 2014 until 2018 (*Data source* ICODATA.IO, https://www.icodata. io/)

sales', have proved extremely popular, and the number of ICOs has increased rapidly since they started in 2014 (see Fig. 2.6).

Regulatory authorities around the world have noted the success of ICOs and are taking steps to ensure they stay within the law. The USA has applied existing legislation to ICOs and associated business activities. Several countries, including Malta, Gibraltar,[17] and Switzerland [44], have all introduced legislation tailored to ICOs designed to protect the market. A recent survey by Deloitte [38] highlighted the threat to the development of blockchain from weak or non-existent regulation. Over-regulation can raise costs for businesses and threaten an entire sector, if not managed properly. But one certainty is that business hates uncertainty; lack of legislation or good regulation brings the fear of unforeseen repercussion. When cryptocurrencies are used for speculation, regulators must act to protect the integrity of markets, to protect consumers and investors, and to safeguard overall financial stability. It is also their duty to prevent illicit use of funds.

---

[17] See http://www.gfsc.gi/dlt (accessed May 10, 2019).

> ## The Dot-Com Bubble
>
> The dot-com bubble was not caused by a lack of regulation, as companies were operating within the laws covering Initial Public Offerings (IPOs). However, the IPOs were often based almost entirely on ideas, with little or no hard evidence of success. Indeed, in some cases they were based on little more than a speculative business plan. Investors were desperate to jump aboard the fast-moving train and become part of a lucrative emerging market. At the peak of the dot-com bubble, the combined market capitalisation of US technology stocks was almost a third of world GDP. When the bubble finally burst, therefore, the impact was severe. Thankfully, the likelihood of a similar cryptocurrency bubble appears unlikely. When peak market capitalisation of cryptocurrencies (including tokens issued by ICOs) was reached in December 2017, it was worth around 1% of the world GDP.

Inevitably, the rise in popularity of cryptocurrencies has led to fear of a repeat of the infamous dot-com bubble, which peaked in 2000 before company values slumped 80% just two years later. As we reflect on the dot-com bubble and its aftermath, we shouldn't forget that some companies, like Amazon and eBay, survived and went on to be huge success stories. Since then, the tech market has matured and the benefits of the new technology will play an integral part in delivering long-term sustainable companies. There are two projects, for example, led by scientific heavyweights that offer an opportunity to deliver significant advances in the technological and business sectors: Algorand[18] and Elixxir.[19] Algorand is led by the Turing Award winner Silvio Micali and Elixxir is headed by David Chaum, a cryptographer who created the first private electronic payment system in the early 1990s [70] and several other privacy-enhancing ideas [69, 71, 74]. Both promise a new generation blockchain, that will overcome the lack of scalability and strong privacy with robust and open technology.

Bitcoin is, in many respects, the engine of cryptocurrencies; it is practical, strong and resilient and with iterations that take it well beyond the original concept. These extensions of the original idea are already well advanced and have produced over two thousand cryptocurrencies and several use cases. We have seen the highs and the lows of both cryptocurrencies and blockchain over the last decade. In the next ten years, we can expect the technology to consolidate its role in the areas where it will provide most benefits. We do not

---

[18] See https://www.algorand.com (accessed May 10, 2019).

[19] See https://elixxir.io (accessed May 10, 2019).

believe the technology will revolutionise every aspect of life, but it will have a significant impact on cross-border payments, in sectors where it is convenient for several parties to work together, like supply chain where several parties need reliable and authoritative documents to trade. We may yet see a currency for the Internet that is independent from all existing national currencies—that would be truly innovative.

3

Cryptography for Busy People

Cryptography! The word alone is enough to send most non-techies diving for cover. And if you also happen to be a time-poor business person with a thousand-and-one important tasks competing for your attention, it would be entirely understandable were you tempted to skip this chapter. However, that would be a mistake that could ultimately cost your organisation money.

Why? Because just as you can't learn a foreign language without having a working grasp of grammar, it's simply not possible to fully understand the exciting possibilities presented by the new world of blockchain technology without having some understanding of the role played by cryptography. If you still need convincing, consider this before considering skipping ahead: all advanced economies, including the UK, rely heavily on computing technology; while most of us might not realise it, cryptography is used extensively to keep our online activity safe. Online banking, Internet shopping, filing tax returns online and some mobile messaging service all rely on cryptography. Because of this, knowing the basics of cryptography can help make you and your business more secure.

Still think it sounds complicated? Well, the good news is that what follows are the principles of cryptography with a minimum of techie and mathematical stuff! Indeed, with a bit of luck, by the time you reach the end of this chapter, you might actually find the topic fascinating.

© The Author(s) 2020
M. G. Vigliotti and H. Jones, *The Executive Guide to Blockchain*,
https://doi.org/10.1007/978-3-030-21107-3_3

Why Cryptography?

To find an answer to that question, let's take the example of Alice, a company CEO, and her colleague Bob, who is head of Corporate Communications. They are going to feature a lot over the next few pages as cryptography gradually becomes clearer. In this instance, Alice and Bob want to meet to discuss the company's annual results, which, although excellent, contain some highly sensitive information that must remain strictly confidential until the next shareholders' meeting. Alice and Bob arrange to meet in the board room, happy in the knowledge that it is soundproof and has even been swept for electronic bugs. They can talk openly, knowing that their conversation will remain secret. However, at the last minute they are forced to abandon their plans and instead have to rely on mobile technology,[1] hardly an unusual way of communicating in the twenty-first century, but far less secure. However, despite initial fears that their conversation might be hacked, they are assured that cryptography will provide the security needed.

Preserving secrecy has been an historical driving force in the development of cryptography. However, modern cryptography is about much more than secrecy: it enables authentication without relying on physical scrutiny; it ensures that sensitive data such as bank statements or company accounts cannot be altered without authorisation; in the case of cryptocurrencies, such as Bitcoin, transactions can be digitally signed, preventing unauthorised access to funds (more examples will be discussed later in the book).

Cryptography,[2] from the ancient Greek 'kryptos', meaning 'hidden' and 'grafein', which means 'written', is the science of designing a secure communication over an insecure channel [78, 87, 92, 102]; some obvious modern applications are email and text-messaging on mobile phones. Secure communication could prevent an *adversary*,[3] i.e. anybody for whom the communication is not intended, from interfering. Methods of interference (see Fig. 3.1) might include:

---

[1] Texts or conversations on mobile phones are not secure, meaning that anybody could intercept the telecommunication signal, and have access to these conversations. Some messaging apps, such as WhatsApp, however, encrypt the messages for security purposes.

[2] Often used as a synonym of the word 'cryptology' which dates back to 1635–45 to the New Latin word 'crypto' and 'logy' meaning *branch of knowledge, science* (see https://www.dictionary.com/browse/cryptology). Strictly speaking, cryptology is used for both 'cryptography' and 'cryptanalysis' the science of identifying ways to flaws into the design of a secure system.

[3] Several words are used in the literature: 'attacker', 'interceptor', 'hacker', 'eavesdropper' or 'enemy'.

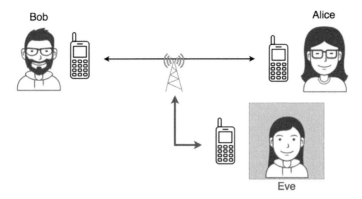

Fig. 3.1 Bob and Alice are communicating over the phone. Eve (she could be an employee, or a competitor, or a member of a hacker group) would like to disrupt communication by: intercepting and listening to the call; or changing part of the message so that either Bob or Alice will receive a different message from the one intended, or interrupting the communication so that the conversation does not happen

1. Listening in on confidential communications[4]
2. Changing the content of communication in a way that, in our example, neither Bob or Alice will notice[5]
3. Blocking their communication altogether.[6]

The relatively young science of modern cryptography has evolved from simple code writing to rigorously devising the security of communication. Modern cryptography has its own language:

Plaintext The message to be hidden
Encryption The operation of hiding plaintext
Cyphertext or Encryption[7] Encrypted plaintext
Cryptographic algorithm A set of finite rules that transform plaintext into cyphertext (*encryption algorithm*) and vice versa (*decryption algorithm*).
Key A random piece of information used by the cryptographic algorithm to transform the plain text into the cyphertext (using an) and vice versa (using the *decryption key*).

To prevent Eve from interfering, as in Fig. 3.1, Bob will insert the text and his encryption key into an encryption algorithm, which scrambles the message and delivers a cyphertext (see Fig. 3.2). Alice will insert the cyphertext and her

---

[4]Violating *confidentiality*.
[5]Violating *integrity*.
[6]Violating *availability*.

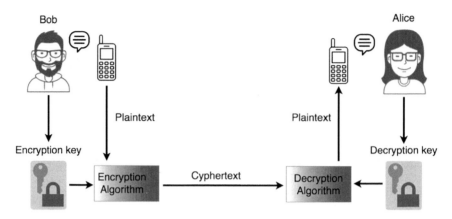

Fig. 3.2 Graphical view of the encryption and the decryption processes. Encryption takes a plain text and an encryption key to output the cypher text; decryption takes the cyphertext and the decryption key to recreate the original plaintext

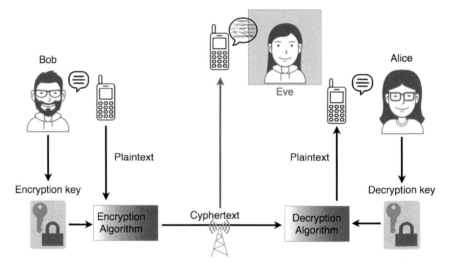

Fig. 3.3 Encrypted phone communication between Bob and Alice, with Eve eavesdropping on the communication

decryption key into a decryption algorithm to retrieve the original message sent by Alice.

How do the encryption and decryption algorithms really work? There are several algorithms [78, 92, 102], yet it is useful to think of the encryption and decryption algorithms as two black boxes: we might not know how they operate, but, like a sausage machine, we can see what goes in and what comes out. These algorithms operate in a similar manner: for the encryption phase, the algorithm takes in both the plaintext and the encryption key and spits out the cyphertext. The decryption algorithm instead takes in the cyphertext and the description key and spits outs the plaintext (see Fig. 3.2).

Why should we trust these 'black boxes' and what does it mean for Alice and Bob to be able to 'communicate securely'? Here, the word 'secure'[8] means that even if the cyphertext is intercepted, a malevolent attacker, like Eve (see Fig. 3.3), would be, in principle, unable to retrieve the original plaintext in a reasonable time frame without the decryption key.

Brute Force Attacks

Simply put, in this context a *brute force attack* means trying every single possible key in turn until the correct key is found. Assuming Alice hasn't shared the decryption key and no attempt has been made to physically coerce it, Eve must:

1. *Obtain* the intercepted (eavesdropped) cyphertext
2. *Attempt to generate* the original key (needed for decryption) used by Bob for the encryption
3. *Decrypt* the cyphertext with the generated key in Step 2
4. *Decide* if the recovered plaintext is acceptable
5. If it isn't, she will need to either generate another key and go back to Step 2, or admit failure.

Taking a worst-case scenario in which Eve has the most powerful computing power at her disposal, what are the chances of her sinister hacking ways succeeding?

Not Enough Time!!
The estimated age of the universe in seconds is 4.32×10^{17}. For Eve, to go through all 2^{256} possible keys, would take longer than the age of the universe, namely 10^{66} seconds (approximately $2^{256}/10^{12}$ at a million times a million keys a second).

A cryptographic key has a length, measured in *binary data*[9] consisting of zeros and ones; conventionally, these are called bits. If we consider algorithms using a 256-bit long key, to successfully carry out a brute force attack Eve might have to go through all possibilities, meaning she can start with a key of 256 zeros, and systematically change each bit starting from the first one until the 256th, a key of

---

[8] Security algorithms can at a later stage be discovered not to be secure. An example was the Data Encryption Standard (DES) [102]. The vulnerabilities of DES were not related to the fact that it was a symmetric algorithm, but to flaws in its design [98].

[9] Binary data is what a computer processes.

256 ones. Eve might therefore need to generate up to 2^{256} keys, which is a very large number.[10] If we make the very optimistic assumption that Eve has one million processors that would be able to test one million decryption keys each second, it might take Eve much, much longer than the current age of the universe to carry out this exhaustive search [92].

Brute Force Attacks in the Art of Codes

Julius Caesar (100–44 BC) developed the Caesar Cypher which requires shifting the letter of each words for three places in the alphabet. The Roman (Latin) alphabet has only 23 letters, so it is relatively easy to see that the plaintext 'Julius' would be encrypted as 'Lxolxv'. Any brute force attack would succeed by shifting back each letter of 'Lxolxv' three places.

A general version of the Caesar cypher, where the number of places to shift each letter is not known in advance, would not guarantee extra security. Let's consider the cyphertext 'Ezi'. A brute force attack requires to systematically shift back each letter a fixed number of places, trying all the 22 possibilities until the original word 'Ave' appears; of course, the assumption here is that the attacker recognises the plaintext. This is where modern cryptography makes a difference. In the Caesar cypher, the number of possibilities to search for the original plain text is simply too small. An attacker can easily exhaust all with brute force. In modern cryptography, it is possible to choose the number of possibilities so that an exhaustive search can be simply too big to be carried out in practice.

Guessing a Key

If the brute force attack seems daunting, Eve could consider the option of merely guessing the encryption key to carry out her evil plan. The odds of Eve randomly guessing the key is the inverse number of all possible keys, i.e. $1/2^{256}$, which is very close to zero. Put another way, her chances of success in this high-stakes guessing game are much, much lower than winning, with one sequence of numbers, the national lottery several weeks in a row.

A *cryptoanalyst* might consider several other possible attacks, including: Eve somehow knowing part of the decryption key, or the decryption key not

---

[10]It is estimated that there are between 10^{78} and 10^{82} atoms in the known, observable universe and 2^{256} is approximately equal to 10^{78}.

being somehow regular,[11] enabling her to increase the chances of guessing the encryption key, or to dramatically reduce the number of possibilities for her brute force attack. But for our purposes, as long as Eve doesn't have access to the decryption key, and Alice and Bob have chosen a secure algorithm, they can relax because their communication will remain secure.

No Need for Security by Obscurity

The knowledge of Caesar cypher enables attacks; hence, it is not secure. It might seem counterintuitive that in modern cryptography the algorithm can be public. It's helpful to think of encryption and decryption algorithms as a sequence of complex mathematical operations that need two numbers, the secret key and the plaintext, to return another number, the cyphertext. Eve can take the cyphertext, she knows the mathematical operations, but she does not know the order in which these mathematical operations must be applied to get the plaintext back. The clever part about modern cryptography is that the order in which these operations are executed is determined by the secret key, and the number of ways in which the operations can be reversed to find the plaintext is so big that it could take the time of the whole universe to try them all.

Staying Secure

In modern cryptography, there are, broadly speaking, two kinds of cryptographic algorithms:

1. *Shared key* or *symmetric*
2. *Public key* or *asymmetric*

The security of the communication is independent from the kind of algorithm: both symmetric and asymmetric algorithms can be secure if appropriately designed.

---

[11] A not randomly generated key can weaken the security of a cryptographic algorithm.

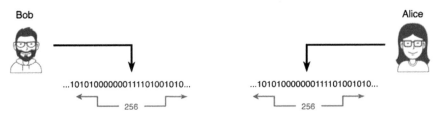

Fig. 3.4 Diagrammatical view of the starting point for secure encryption between Bob and Alice, Bob and Alice having exchanged the same key for encryption and decryption

Shared Key Cryptography

To better understand how symmetric algorithms work, let's assume Bob wants to send the message 'Hello World' to Alice. In shared key cryptography, Bob and Alice will already have a shared secret: this is the *secret key*, which they keep to themselves and is used both for encrypting and decrypting their communication. In other words, the encryption and the decryption key described in Fig. 3.2 are now, in fact, the same. Alice and Bob could use the same key for a long time, or change the shared key every time a new message is sent; this makes no difference to the algorithm or to its security so long as none of the shared keys[12] is in Eve's possession or of anybody else. Together with the key, Alice and Bob have also agreed on a specific algorithm to use, which in turn could determine the length of the key (see Fig. 3.4). The next step is for Bob to encrypt the message 'Hello World' by using both the chosen algorithm and the shared key, and to send the result, a cyphertext, to Alice (see Fig. 3.5).

We know that Eve's chances of guessing the secret key or stumbling across it are negligible. But might her chances be improved by knowing the algorithm used by Bob and Alice? Luckily, modern cryptography does not rely on the secrecy[13] of the encryption or the decryption algorithm. In fact, it's considered best practice to make the algorithm known to the community and place the implementation code in the public domain so that both can be scrutinised for errors by the cryptographic community.[14] Remember that cryptographic algorithms can be secure, but errors in the implementation code can undermine the theoretical security (Fig. 3.6).

---

[12]A secret key would be considered compromised and needs to be changed, even if only part of the key is in possession of Eve. In practical cryptography, key management requires to change key regularly to mitigate the risk keys becoming compromised.

[13]This concept, also known as 'Security by Obscurity', was first formulated by Kerckhoffs in 1883 as part of his six design-principles of cryptosystems [90].

[14]Making an algorithm and the associated software publicly available does not make it either automatically more secure or trustworthy. Open-source software can become more reliable if there is a large community of developers that uses, tests and improves the codebase.

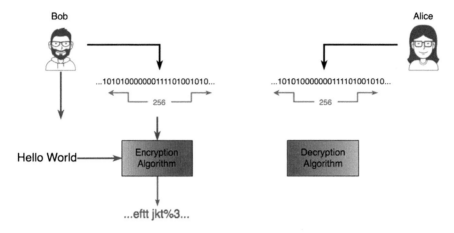

Fig. 3.5 Bob has encrypted 'Hello World' using the shared key and the agreed algorithm

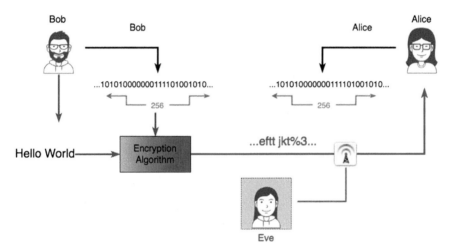

Fig. 3.6 Bob sends the cyphertext to Alice over an unsecured channel and Eve can retrieve the cyphertext

Assuming Bob and Alice have kept their shared secret keys secure, they will be the only people able to retrieve the original message; hence, the communication is completed when Alice takes the cyphertext delivered by Bob, and uses the secret key, together with the decryption algorithm to retrieve 'Hello World' (see Fig. 3.7).

There are two issues with shared key cryptography:

Secure exchange of the key Prior to starting any communication, Bob and Alice need to agree what key to use. The security of all their messages depends on them securely sharing the key. They can do this in several ways: they could,

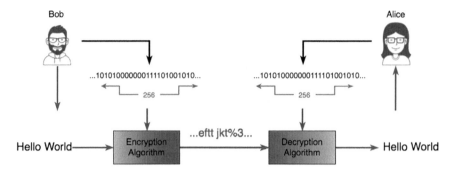

Fig. 3.7 Graphical representation of the steps necessary for Bob to send an encrypted message to Alice, and for Alice to decrypt it

for example, physically meet and exchange the key—sending the key via email or via courier introduces the risk of third-party access. Securely sharing a secret key between two parties can be challenging. In cryptography, there is a whole area dedicated to key management that includes: key creation, distribution, storage, destruction and escrow.

Pair communication If Alice wishes to secretly communicate with more individuals or even a group, it's not practical to have a single shared key. For security, Alice will need to hold as many keys as the number of people she wants to communicate with, and communicating with a large group will result in Alice having to manage and secure several keys.

So far, we have covered symmetric or shared key cryptographic algorithms, where sender and receiver have access to a common secret key. But when applied to large groups of people, they become cumbersome. Asymmetric, or public key, cryptography, as discussed below, overcomes both issues.

Public Key Cryptography

In public key cryptography, two different keys are required: one for encrypting the message and a different key for decrypting the message. Let's return to our example of Bob who wants to send the message 'Hello World' to Alice (see Fig. 3.8). This time Alice has two keys:

1. A *private key*
2. A *public key*.

Alice's public key can be put in the public domain—via social media, a webpage or as part of an email footer—to enable anybody, including Bob,

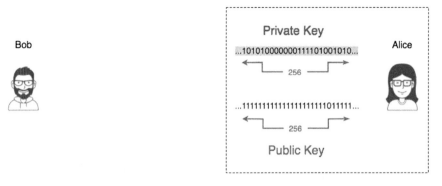

Fig. 3.8 Graphical view of the keys held by Alice in asymmetric key cryptography, when Bob wishes to initiate the communication

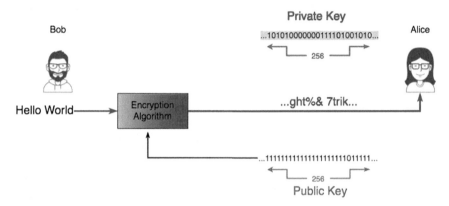

Fig. 3.9 Bob encrypts the message 'Hello World' using Alice's public key

to communicate with her. It doesn't matter where Alice publishes her public key, so long as she protects her private key. To communicate securely, Bob takes Alice's public key to encrypt the message 'Hello World' and sends the cyphertext to Alice (see Fig. 3.9). Now Alice has to use her private key to decode the cyphertext and retrieve Bob's original message (see Fig. 3.10).

The public and the private keys are paired together in a unique way: for each private key, there exists only one public key and vice versa. As a result, it's not possible to decode an encrypted message with an alternative key. As long as Alice doesn't share her private key with anyone else, she will be the only person who is able to decode Bob's scrambled message.

There is something rather astonishing about public key cryptography. It is, at least to some extent, intuitive that if Alice and Bob share a secret, as in symmetric cryptography, the attacker, Eve, needs to find that secret. In asymmetric cryptography, Eve also knows the public key, and the question is: Can she use the public key to get the private key?

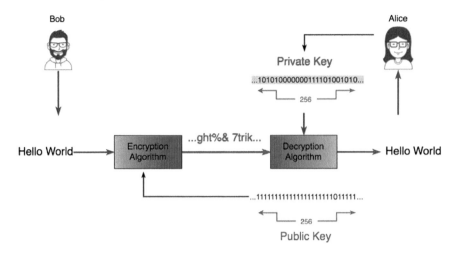

Fig. 3.10 Alice takes the cyphertext produced by Bob, used her private key and retrieves the plaintext, 'Hello World'

Thankfully, the answer is 'no',[15] and the reason why is firmly rooted in mathematics. Now any message, however long, consists of 0s and 1s and thereby can be seen as a (binary) number. The public key is a number as well and encryption consists, essentially, of using these two numbers within a complex mathematical formula to generate a third, which then represents the encrypted message. This process is sufficiently complex to make it very difficult to reverse, even with the public key in hand. The private key, however, can uniquely reverse the encryption process but to ensure robustness the private key must be long and difficult to guess.

Public key cryptography creates trust between sender(s) and receiver. Before public key cryptography, it was believed that secure communication had to rely on secret key(s) that needed to be exchanged between parties who already knew each other. In public key cryptography, Alice does not need to know Bob: she generates the key pair by herself and publishes the public one. Bob does the rest as far as encryption is concerned, and Alice will do her part at the decryption stage.

Encrypting using public key algorithms normally takes much more time than symmetric key algorithms. So public key algorithms are used to securely share secret keys. In other words, if Alice and Bob want to communicate securely using a symmetric algorithm, they can both generate their key pair and follow series of steps [78] to achieve a secure exchange of keys.

---

[15] Provided the private key is sufficiently long.

Digital Signatures

Because of the unique pairing between private and public keys, public key cryptography is also used as a proof of signature; for example, Alice, as CEO, can use her private key to securely sign a contract. Now let's consider an example where Bob and Alice are at different locations, and Bob needs Alice to sign a contract for an important client. One option would be for Alice to scan her signature and add it to the contract. This could constitute an *electronic signature* [26]. For the client, there exists the risk that Alice is not actually committing to the contract: if her company cannot deliver, she could claim that somebody else scanned her signature and added it to the contract. In other words, she can *repudiate* the signature and cast doubt on the legal validity of the contract.

> ### Public Key Cryptography—A Very British Invention!
>
> Public key cryptography has been deemed to be worth of the highest scientific prize in computer science: the Turing Award. Three MIT researchers, L. Adleman, R. Rivest, and A. Shamir won this award in 2002 based on the invention of the RSA algorithm in 1978 [100].
>
> It is not well known that five years earlier, in 1973, the British mathematician Clifford Cocks had already invented the RSA. Cocks was working for the British Government Communications Headquarters (GCHQ); hence, his work was classified as secret and not released until 1997, twenty years after Rivest, Adleman, and Shamir had published their independent discovery.

By using *digital signatures* via public key cryptography, this is no longer possible. As a starting point, we have a contract in digital format that Alice would like to send to the client. Alice is in possession of both the private and public key, and it is assumed that nobody would dispute that the public key (see Fig. 3.11) exclusively belongs to Alice. Alice will now apply the encryption algorithm to contract[16] using her private key, to produce a scrambled message (see Fig. 3.12). By contrast, in the previous two secret communications, Bob was using Alice's public key: but the role of the key and the party receiving, the information has changed.

Receiving a scrambled document, the client will still need to be sure that the contract has actually been signed by Alice. The key point here is that the client does not need to go back to Alice to ask for more evidence. They can just take Alice's public key and decode the scrambled message (see Fig. 3.13).

---

[16]Strictly speaking Alice will apply the encryption algorithm to the cryptographic digest (see Chapter 4, section 'How Does the PoW Work?') of the contract.

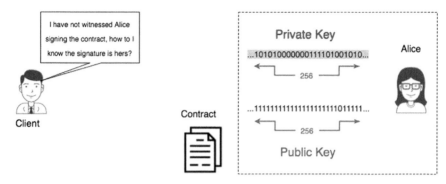

Fig. 3.11 Starting point for the digital signature

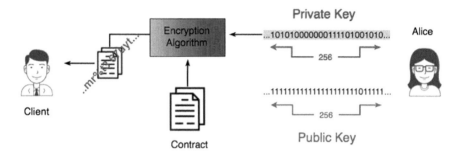

Fig. 3.12 Alice signs her contract using her private key, instead of the public key, and sends it to the Client. By contrast, in the previous example about encryption, Bob, an external party, used Alice's public key

If the client retrieves the original contract (see Fig. 3.13), then that is proof that Alice, and only Alice, has signed the contract—as the private and the public keys are uniquely paired. If Alice had used a different private key, or somebody else had scrambled it pretending to be Alice by using a different key, then the client would not have succeeded in retrieving the original document with Alice's public key.

In this example, no consideration has been given to the secrecy of the contract. Eve could intercept the cyphertext, and given that Alice's public key is meant to be known to the wider public, Eve can read the contract (see Fig. 3.14). In real life, the secrecy of the contract could be crucial, and other cryptographic measures would need to be taken to ensure it.

Finally, how can we guarantee that Alice's public key really belongs to her? Wouldn't it be possible for Eve to impersonate Alice, by telling everyone that her public key belongs to Alice? This is where *Certificate Authorities* (CAs) come into play; these are independent third parties that associate public keys to authenticated individuals or organisations by issuing a digital certificate.

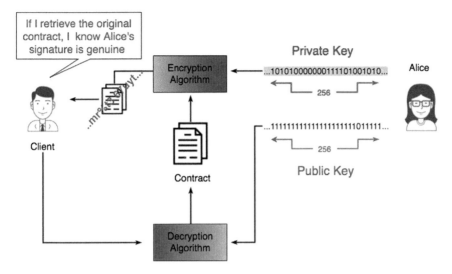

Fig. 3.13   The verification process takes Alice's public key and derives the original contract

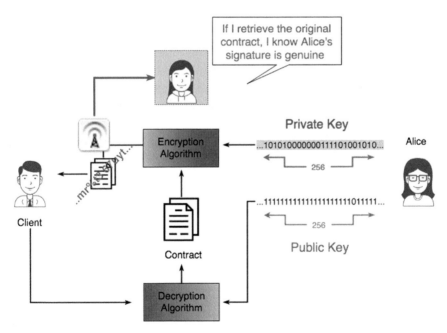

Fig. 3.14   Eve can intercept the scrambled contract and with Alice's public key can decypher and read the contract

Of course, certificate authorities add an extra level of trust, but they are not needed: it suffices to be able to authenticate the individuals behind a public key. Digital certificates are widely used for secure communication on the Internet. For example, how do we know that the site https://www.gov.uk is not a spurious site set up by criminals trying to steal information? The simple answer is the digital certificate, issued by a certificate authority (CA), which is automatically read by the browser when accessing the website.[17]

How Is This Relevant to Blockchain?

Public key cryptography is an essential part of blockchain technology that enables users to:

- Digitally sign transactions that transfer funds or data to authenticate the sender.
- Prevent fraudsters from transferring funds or data on behalf of the legitimate account holder.
- Ensure the integrity of the data on the blockchain by preventing anybody from changing it.

In the next section, we will find out more about how cryptography is a core part of blockchain: the public key acts as an identifier to hold assets; the private key is kept secret and is used to sign a digital transaction to make sure the origin of the transaction is legitimate.

Typically, in a public blockchain the ledger is public and anybody can view the content. In some advanced blockchains where the privacy of the users is crucial, advanced cryptographic techniques are used to prevent third parties from detecting movements of funds to specific addresses. In the so-called permissioned blockchains, cryptography is also often used to ensure an extra layer of confidentiality when the users would like to communicate about matters that do not concern other participants.

Summary

Cryptography is the science of making communications secure even in the presence of powerful adversaries. It is powered by mathematical reasoning, making it possible to quantify the security of algorithms. Symmetric key cryp-

---

[17]Accessing a website with an incorrect certificate will normally cause the browser to give a warning.

tography uses just one key for encryption and decryption, and it presents two challenges: securely sharing the secret key with the other party and managing as many keys as there are communicating parties. Public key cryptography is said to be asymmetric as the key used to encrypt is different from the key used to decrypt. It overcomes the issues presented by symmetric key encryption, as one party, say Alice, generates the private/public key pair and publishes the public key enabling everyone to send encrypted messages.

Private keys can also be used to sign messages which, using the public key, can be used to validate the authenticity of the sender. While blockchain technology makes good use of cryptography, cryptography has been deployed to secure aspects of our society for long time.

Designing new cryptographic algorithms or implementing old ones in software are both very difficult tasks that require highly specialised skills. Small, undetected mistakes can have disastrous consequences. Blockchain technology uses mostly public key cryptography, alongside other interesting clever functions which play a crucial role in updating the ledger.

Frequently Asked Questions

What does it mean for a key to be 256-bit long?

Cryptographic keys are measured in binary data made up of zeros and ones. A 4-bit long key will four zeros or ones; examples of a 4-bit long key are '0000' or '1000' or '1010'. There are another 29 different keys that can be written down for a total of $2^4 = 32$. If a key is 256-bit, there are two hundred and fifty-six zeros and ones. Nobody can create 2^{256} keys, that's just too big a number!

How should I store a private key?

Secure storage of private keys should be taken very seriously [92]; below you will find are some comments, not to be taken as a guideline. For organisations, there are standards that support secure storage, for example [62]. Make sure you are in control of your private keys: you need to know how many copies you have made and where they are stored, and who has access to them. Never store your private key somewhere where it can be easily accessed: if your laptop is connected to the Internet, then malware can copy the key, a piece of carelessly left around data can be easily spotted and copied. Keep the private key encrypted, on one or more hardware disks not connected to the Internet. If you have more than one copy of the private key, make sure the two copies are physically located in different places, and make sure that you have regular access to all of them.

If cryptography is so secure and widely used, how come newspapers and crypto news websites report security attacks?

There are only a few security attacks in the world that involve breaking an algorithm or retrieving a private key via cryptoanalysis alone. Security attacks are often due to the weakness of other parts of the IT system. One of the more infamous attacks associated to storage of the private keys is Mt. Gox, one of the biggest crypto-exchanges in the world back in 2011. Mt. Gox started in 2010 and closed down, filing for bankruptcy, in 2014. Poor key storage leads to a loss of 744,408 Bitcoin private keys belonging to customers; the hacker did not attempt to generate or guess any of the private keys. Custody of the private key is crucial to ensure the security of any system.

Do encryption and decryption keys come in different lengths?

Yes. In public key algorithms, where the two keys are different, the private and public key pairs can have different lengths.

Could you provide examples of public key cryptographic algorithms used in real life?

Probably the best-known public key cryptographic algorithm is RSA. RSA is used ubiquitously over the Internet and is one of the oldest public key algorithms. Among other recognised algorithms is the Elliptic Curve Digital Signature Algorithm (ECDSA) with a 256-bit long key, which is used in the Bitcoin blockchain and in several others.

Could you provide examples of symmetric key cryptographic algorithms used in real life?

An example of symmetric key algorithms is Advanced Encryption Standard (AES) widely used in the encryption of hardware, file transfers and for several other activities. It is a flexible algorithm that enables different key lengths: 256 bit and 512 bit. More information can be found in [92, 102].

Should I share my private key with people I trust, if I want them to act on my behalf?

Just as you shouldn't share pin numbers or passwords, you shouldn't share a private key.

Can I sign a document using the key from a symmetric key algorithm?

No. A digital signature needs to be verified by third parties, without undermining the security of the encryption schema. Because of the role of the public key in the verification process, this is only possible with public key cryptography.

4

Bitcoin and Blockchain: The Fundamentals

Bitcoin is open-source; its design is public, nobody owns or controls Bitcoin and everyone can take part [28].

Bitcoin, a trailblazing cryptocurrency and the first practical deployment of a blockchain, was established over a decade ago. Since then, it has become the most capitalised cryptocurrency out of approximately two thousand in existence and is arguably the one with the highest name recognition.

This chapter examines Bitcoin's success and explains why it is one of the most trusted cryptocurrencies in the world today.

What Is Blockchain Technology?

A blockchain, in a nutshell, is a *data-recording* platform that makes it safer for businesses or individuals to work together by creating.[1] Trust is created by permanently attaching, in a strict chronological order, information or data according to a set of pre-agreed rules that are automatically enforced. The data is typically bundled together to form a *Block*. Each newly formed block is linked to the previous block forming a chronological chain called the *ledger*: no human effort is necessary for participants to agree on the status of the ledger.

---

[1] Trust is defined here as: 'Freedom from suspicion or doubt'.

© The Author(s) 2020
M. G. Vigliotti and H. Jones, *The Executive Guide to Blockchain*,
https://doi.org/10.1007/978-3-030-21107-3_4

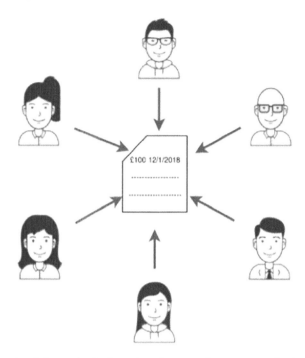

Fig. 4.1 Example of chronologically ordered transactions managed using an automatically enforced consensus mechanism

The link between blocks is secured by cryptographic means, hence losing or altering any of the data inside any of the blocks is (almost) impossible: if one block is changed, accidentally or maliciously, all subsequently linked blocks need to be changed to meet the pre-agreed rules that determine how data is added to the ledger. Unlike other data recoding platforms (typically referred as a), the does not belong to a single entity, but is maintained and curated by a group of participants, small or large, all of whom typically have the same rights and privileges. Just for the sheer number of participants, everyone can trust that everyone else will follow the rules. Breaking the rules by collusion of parties is (nearly) impossible. The blockchain puts every single participant on a level playing field (Fig. 4.1).

To see how blockchain works, let's take the example of a cloud-based data-recording platform, where a group of businesses share 'sensitive data' and a budget of £100. The businesses, or participants, agree the following rules which are encoded and enforced by the platform:

1. Each business needs to login and record an expense request.
2. Expenses are approved only if every single party agrees.

3. Every business can see who has approved the expenses—and who has not!
4. Once the budget runs out, participants contribute equally to a new sum of £100
5. Chronological order is preserved when new data, such as expenses requests and approvals, is added.
6. Recorded data cannot be deleted.
7. Data history is visible to everyone.

Let's assume data is securely encrypted, participants are able to read and write each other's data and enjoy the same rights and privileges. This system has some significant advantages:

Transparency All participants can see what everyone else is doing in real time.
Data history All participants can see all the transactions.
Digitalisation Being cloud-based, there's no need for lengthy paper-based processes.
Efficiency A simple login provides access to each participant.
Completeness All data is stored on the cloud and expenses can only be agreed on the system.

So far, so good. But what if the cloud provider, also called the *super-participant*,[2] acts to change data, deletes history or even stops providing the service? Normally, businesses would protect themselves from a situation like this by signing a contract preventing interference by the cloud provider. In contrast to blockchain, there is nothing to fear from a cloud provider: blockchain participants rely on technology that negates the need for a super-participant and delivers independent verification that rules are being adhered to. Put simply, blockchain technology, if appropriately used, makes it extremely difficult for participants to cheat.

Privacy, Anonymity and the Double-Spending Problem

Let's look at the evolution of Bitcoin and cryptocurrencies in more detail.
All of us will have handled cash at some point in our lives. Cash transactions are as follows:

---

[2]In blockchain jargon, the 'super-user' or 'super-participant' described in this example is generally called a 'third party'.

Anonymous Cash can be exchanged without knowing the people involved in the transaction.

Atomic There is no need for a third party or an intermediary.

Irreversible Once cash is moved from one person to another, a third party can't intervene to reverse the transaction.[3]

Free Normally, there's no fee for cash transactions.

There are clear advantages of cash, but we are also very much aware of its drawbacks: it wears out easily; it can be forged; it can be lost or stolen; it is used in illegal transactions; it can only be exchanged by people in the same location; and it is impractical for large purchases, such as a house. The introduction of credit/debit cards and Internet banking changed the rules of the game and has led to a dramatic reduction in the use of cash, but with the direct consequence that atomicity and anonymity of the transactions no longer hold. There are different reasons as to why this is the case.

Atomicity for electronic transactions, so to pay somebody directly electronically (whether in a private manner or not) is not possible: as research has shown in the 90s, there is always need of a trusted third party or bank to prevent double spending, i.e. spending the same coin twice [70]. Any monetary system that enables double spending is simply worthless; of course, spending fiat cash, like pounds or euros, twice is physically impossible. In the digital world, i.e. where money is processed on computers, the bank prevents double spending by checking that when Alice wants to move the money into Bob's account, the money is taken away from Alice's balance and is inserted only into Bob's account.[4]

For financial institutions who offer credit to their customers, as credit card companies do, knowing the details of the identity of customers is necessary to hedge the risk of default and complying to money laundering legislation. As credit cards are an increasingly popular means of payment, it has become apparent that banks and financial institutions could end up having a significant amount of data on the spending habits of private citizens. Keeping track of Bob's spending habits can be very revealing and gives a very accurate and detailed idea about his live: it would reveal when Bob is away from home and for how long; which places he visits and when he is abroad; his daily routine; knowing which books Bob buys could reveal his politically inclination; and so on. It is simply not possible to acquire all these data through cash transactions.

---

[3]We are not considering criminal activities where cash could be returned in illegal ways.

[4]Ultimately, the banks reconcile their accounts with the central bank, which in the case of the UK is the Bank of England.

Privacy and anonymity are strongly related; the prospect of financial institutions accumulating personal data creates concerns on the risk of privacy violation. If transactions cannot identify the person that is behind it, then the privacy of that individual is also protected—at least as far as the financial transactions are concerned. Hence, producing a digital payment system that prevents double spending enables individuals to remain anonymous and thereby protects their privacy, and does not need a third party would be making some significant technical advances. In the 90s, academics and industrial researchers started to consider how to enable digital transactions that simulate cash transactions, i.e. are atomic, irreversible and privacy-protecting [25, 70, 72]. Bitcoin is that payment system.

There's nothing mystical about Bitcoin, which deploys well-known cryptographic techniques first developed more than thirty years ago, and built on well-known academic ideas [25, 61, 83, 85, 96]. Bitcoin's indisputable innovation is to eliminate the need of a third party that keeps track of all transactions to prevent double spending by:

- Maintaining only one public store record that is the same for all participants—the ledger.
- Involving the community to update the ledger via a pre-defined set of rules.

Interestingly, Bitcoin operates in an *adversarial environment*, meaning that it is set up to protect itself against participants who are dishonest and would:

- Attempt to steal bitcoins.
- Stop the Bitcoin network from operating.
- Double spend their coins.

Bitcoin has been designed to be resilient to the attacks described above, and users can securely send and receive bitcoins. In a nutshell, Bitcoin enables participants to

- Send bitcoins to each other.
- Maintain the integrity of the ledger.

Sending bitcoins is anonymous in that it doesn't require participants to disclose personal data. Instead, to transfer a bitcoin to a user all that is needed is their *Public address* which is effectively derived by the public key.[5]

---

[5]The public keys are shortened in such a way that it is highly improbable that two different keys, belonging to two different users, will yield the same address (see section 'How Does the PoW Work?'), about cryptographic digest.

The supply of Bitcoins is finite, with a limit of 21 million set at the time of its creation—why this figure was chosen is not clear.[6] Not all bitcoins are yet in circulation although bitcoins are minted every day through a process of *mining*. At the time of writing[7] well over 17 million have been mined and are currently in circulation.

Bitcoin's price is driven by supply and demand. While Bitcoin has a rudimentary monetary policy, there is no central bank or authority that controls it or decides the exchange rates. An algorithm determines how many bitcoins are released to the public. Significantly, there is no national economy underpinning the value of each bitcoin.

How Does Bitcoin Work?

To understand how Bitcoin works, let's go back to Bob and Alice, whom we met in Chapter 3. Alice now wants to send some bitcoins to Bob. We'll assume she knows Bob's public address as well as her own. Because public and private keys are uniquely paired, both Bob and Alice hold a private key as well.

Alice has a balance of 10 bitcoin and wants to send 1 bitcoin to Bob. Bob has a balance of 7 bitcoin (see Fig. 4.2). Alice can't rely on one single party to make the transaction. She will have a piece of software, running on her mobile or laptop, effectively running a *node in the Bitcoin network*. To initiate the transaction, Alice needs to specify:

1. Bob's public keys, which are the *recipient's coordinates*
2. The amount of bitcoin Alice wants to send, which has to be smaller than the amount she currently holds
3. The amount of fees Alice is prepared to pay to ensure that the transaction takes place.

Alice signs the transaction with her private key and submits it to her node, which forwards the transaction to other nodes in the network. By signing the transaction, Alice shows she is the legitimate account owner. Perhaps surprisingly, *Bitcoin transaction fees* are determined by the user, a strikingly different arrangement from conventional transactions. The value of the fee normally determines the speed of the payment: the higher the fee, the quicker the transaction will be settled.

---

[6]See, for example, https://en.bitcoin.it/wiki/Controlled_supply (accessed September 10, 2018).
[7]June 30, 2019.

Fig. 4.2 Bob and Alice with their private keys and public addresses. The Bitcoin ledger shows the bitcoin balance for each of them. Addresses and private keys are uniquely paired

How Are Transactions Processed?

Alice can't update the Bitcoin ledger by herself; otherwise, she would be able to spend bitcoins twice. Sitting between Alice and Bob are miners (see Fig. 4.3), commercial enterprises that run a node of the network, who have a copy of the Bitcoin ledger and continuously collect transactions. Each miner can choose from a pool of transactions which to process, and bundle the chosen transactions together in a *block* of a fixed size.

Understanding Bank Payments

To use the existing banking system, Alice can send £1 by providing Bob's bank details to her bank. To execute the transaction Alice's bank will send an instruction via a secure payments platform allowing the transaction to be processed; Alice's bank will decrease the balance on her account by £1 and Bob's bank will increase Bob's balance when instructed to so. The two banks are changing their own private ledgers. By contrast, in Bitcoin there is only one ledger, available to all users (see section 'How Does Bitcoin Work?').

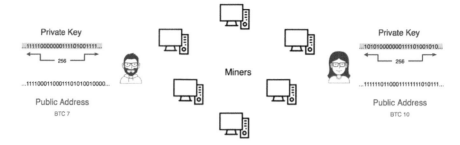

Fig. 4.3 Bitcoin and two crucial group of participants: the users and the miners

What Are Miners?

In the Bitcoin jargon, mining refers to the process of attaching a new block to the Bitcoin ledger. Anybody participating into this process is called a *miner*. Miners are typically commercial enterprises that operate for profit, as they are rewarded with newly minted bitcoins as well. They are essential to ensure that all bitcoins are produced. As mining has been deliberately designed to be resource-intensive, it shows similarity with the extraction process of commodities such as gold or diamonds. There was a time when it was possible for individuals to mine bitcoins with their own computing equipment; whilst still technically possible, the economics of doing so no longer make sense.

The goal of the miner is to attach the newly formed block to the ledger. Bitcoin protocol makes it difficult for a new block to be attached; to conform to the rules, new blocks must contain the answer to a cryptographic puzzle called *Proof of Work* (PoW).

PoW is solved by trial and error; this process can't be accelerated by using clever solutions or algorithms and is very similar to making an exhaustive search for a private key (see Chapter 3). For example, even using the latest version of AI would not provide any advantage to miners.[8] It requires a significant computational effort, meaning numerous computers working without interruption. Verification of the solution of the puzzle is, on the other hand, quick and energy efficient.

Broadly speaking, Bitcoin has two groups of participants: the users who deploy the network as a payment system, and the miners who collect transactions and securely update the ledger. Miners compete against one

---

[8]It is accepted in the academic community that the only way to solve the PoW is via trial and error; however, this has not been mathematically proven yet.

another to solve, as fast as possible, the PoW and the successful miner attaches the new block. Other participants in the network verify that the PoW is correct, and then, the race to solve the next PoW and attach a new block restarts, and so on. A new block can be attached on average every 10 minutes. Miners individually select transactions from the shared pool for their own blocks, and typically two blocks could have a few common transactions.

If the winning block contains Alice's transaction, then Alice's and Bob's balances are considered updated. The miners then move to the next block, gathering newly available transactions or old transactions that didn't make the latest block.

How Do Miners Make a Living?

Every time a new block is attached to the ledger, the successful miner is rewarded, the reward being divided into two parts:

1. Newly minted bitcoins
2. Collections of all fees for each transaction contained in the latest block attached.

These rewards change over time. Rewards relating to transaction fees depend on the number of transactions in a block, and what fees users like Alice are prepared to pay. Miners have the right to choose their transactions using whatever policy they deem successful.

Time	Reward
2009	50 BTC
2012	12.5 BTC
2016	12.5 BTC
Expected Time	Reward
May 2020	6.25 BTC

A more substantial reward comes from newly minted bitcoins; at the time of writing, a miner gets rewarded about 12.5 bitcoin for every newly created block that gets attached to the blockchain and stays attached. This reward is not fixed, but changes with every 210,000 blocks. The table on the side of the page shows the newly minted bitcoin reward since Bitcoin started existing in 2009 with the genesis block.

At time of writing well over 583,000 blocks[9] have been mined, with miners spending a significant amount of money on electricity and specialised hardware to be the first to solve the PoWs and gain the rewards. As with any business,

---

[9]A web-display of the Bitcoin ledger can be found at https://www.blockchain.com/en/explorer, https://blockchair.com/bitcoin or the reader can download the Bitcoin Core https://bitcoin.org/en/download and have a copy of the ledger on a laptop or a desktop. Either of them can help to determine how many blocks have been mined.

revenues must exceed costs; on average, 144 blocks a day are attached to the ledger, and hence, the daily revenue is easily computed by considering

$$\text{Daily Revenue} = 144 \times \text{current reward} \times \text{bitcoin price}$$

Considering that, at the time of writing, bitcoin trades at approximately $10,100, the miners' daily revenue is

$$\underbrace{\$18,180,000}_{\text{Daily Revenue}} = \underbrace{144}_{\#\,blocks} \times \underbrace{12.5}_{\text{current reward}} \times \underbrace{\$10,100}_{\text{bitcoin price}}$$

The World's Most Expensive Pizzas!

It took more than a year for Bitcoin to become established. On May 22, 2010, Lazlo Haneyecz[a] bought two pizzas, costing $41 in total, for 10,000 bitcoin. At the time of writing, 1 bitcoin trades for $10,100 which makes Haneyecz's pizzas probably the most expensive in history. It's astonishing that a cryptocurrency, without a fully-fledged organisation behind it, with no investment, and operated by volunteers, has acquired so much interest, and risen so dramatically in value.

_____

[a]For the detailed historical recount of this story see https://en.bitcoin.it/wiki/Laszlo_Hanyecz (accessed September 10, 2018).

The daily revenue is distributed among the miners that successfully have attached blocks. Figure 4.4 provides an indication of the average fee distribution in recent times.

This revenue stream doesn't take the transaction fees into account, which are another way to incentivise the miners to keep the bitcoin ledger growing. The market price for transaction fees is determined by demand: the higher the demand to insert transactions, the higher the fees. Users can make the choice to pay above or below the market price, depending on how quickly they want their transaction to be processed. High-fee transactions are attractive to miners and as a result are processed faster; transactions with no fees are processed simply because miners have an interest in filling blocks and obtaining newly minted coins.

If miners are crucial to maintain Bitcoin, and their survival depends on the bitcoin reward, what is the future for Bitcoin? Nobody can be sure what will happen to bitcoin when the ceiling of 21 million coins is reached sometime in the next century—the current estimated date is May 7, 2140 (see https://en.bitcoin.it/wiki/Controlled_supply). When all bitcoins will be minted, miners' profitability will be dependent on transaction fees only, meaning energy prices falling or fees rising.

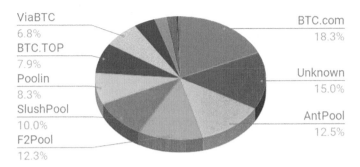

Fig. 4.4 Bitcoin miners' computational power distribution (hashrate) in 2019. Some miners make themselves known, and others prefer to remain anonymous. The graph can be used as an indication of the average distribution of the reward fees (*Source* Blockchair https://blockchair.com/bitcoin, accessed May 22, 2019)

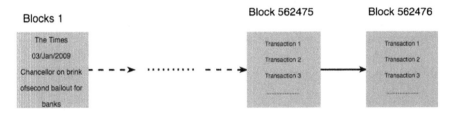

Fig. 4.5 Abstract view of the Bitcoin ledger, which accounts for all transactions and all blocks since the genesis block on January 3, 2009, until the latest block, 562476, under examination

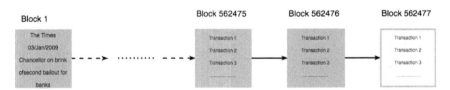

Fig. 4.6 The new block in green 562477 is attached to the Bitcoin ledger by an *honest miner*

The Case for PoW

Alice, or any other user, can't attach a new block without supervision. However, what is less clear is why miners themselves can't choose a block and attach it to the ledger (see Figs. 4.5 and 4.6). Firstly, as we saw earlier, Bitcoin assumes that everybody is trying to cheat! In theory, a miner could collude with Alice and spend the coin twice on her behalf. The PoW ensures that miners can't acquire discriminatory powers; in other words, the PoW is a mechanism that serves two main purposes:

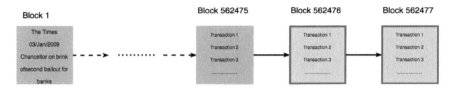

Fig. 4.7 The dishonest miner needs to computer two PoW for both blocks in the same amount of time that other miners will compute the PoW for one block

1. It randomly selects the next block and the miner that attaches it.
2. It maintains the integrity of the ledger, preventing overriding of past transactions.

Ensuring miners are randomly picked prevents collusion and power from being consolidated in the hands of a few parties. Miners can only increase their chances of solving the PoW by using more computational resources. That's similar to buying multiple lottery tickets to only marginally increase the chances of winning. Every miner has access to the same computational resources, which are commercially available worldwide, and hence, everyone has the same opportunities. The PoW ensures a level playing field: no miner is favoured in attaching new blocks, and no miner can grab more power or earn more without spending more money.

To show how this works, let's imagine a miner wants to cheat the system and change block 562476 while the other miners are currently computing the PoW for 562477 (see Fig. 4.6). Under the rules, miners must deliver a valid PoW for each block. The ledger is built in such a way that each block depends on information stored in the previous one; a cheating miner would need to solve the PoW for two blocks 562476 and 562477 (see Fig. 4.7) in the same amount of time that all other miners are computing the PoW for 562477 only. Efforts to cheat the system this way are destined to fail as it's simply impossible for a miner to provide the PoW for two blocks in time.

PoW also acts as an economic disincentive against, for example, a *Denial of Service attack* (DoS). Any attacker seeking to attach their own version of a block will need to use a substantial amount of energy just to attempt to calculate the solution of the PoW. Attackers would have to take a commercial view on how many failing attacks can be financed, as PoW costs money to produce in the first place. The unsuccessful attacker will have invested a substantial amount of money in looking for the solution of the PoW without reaping the rewards.

How Does the PoW Work?

Each block is attached to the next one, and any change in a block will require changes in all subsequent blocks. Why? Miners are required to compute a number, called a *nonce* that effectively connects one block to the next, creating a *secure chain of blocks*. To consider this in more detail, let's look at the example of a *machine* that, given an input, produces a 256 bit-long output; we call this machine **SHA 256** (see Fig. 4.8).

As an input to SHA 256, we could give the text 'Hello World' or the digital version of 'War and Peace', and in both cases, we will receive back two different strings of 256 zeros and ones. This machine is smart in the sense that [95, 108]:

- It delivers the 256-bit output in a reasonable amount of time (not like the amount of time it takes an attacker to guess a private key).
- From the 256-bit output, it is not possible to guess the original input (provided that the original input was sufficiently complex to start with).
- Chosen an input, it is extremely unlikely that anybody can find a different input that delivers the same 256-bit output.

In computing jargon, the output is a *cryptographic hash*.[10] Why is this relevant? SHA 256 is at the core of connecting the blocks. Let's think how

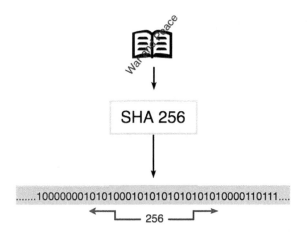

Fig. 4.8 Mental image of the SHA 256. This is a black box machine meaning we cannot look inside and find out how it works, we can only observe the input and output. Regardless of the input, the SHA 256 will return a string of 256 zeros and ones

---

[10]Sometimes it is also called indexdigital fingerprint *digital fingerprint* or *cryptographic digest* [78].

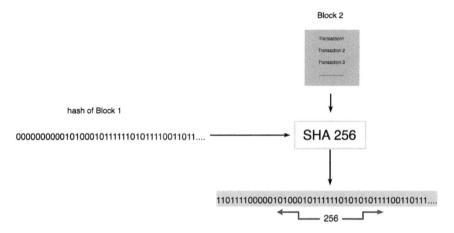

Fig. 4.9 First 'attempt' to link block 1 and block 2. This would not work, as miners can create arbitrary blocks with the chosen transactions

we can connect the blocks from the genesis block. The first attempt is to get the cryptographic hash cryptographic hash of block 1 and all the transactions collected to form block 2, and insert these two inputs in the machine (see Fig. 4.9). The result would be another cryptographic 256-bit hash that will be inserted in block 3, and so on. This leads to a chain of blocks (see Fig. 4.9).

Of course, all miners are creating different blocks, but only one block can be attached to prevent double spending. If two blocks could be attached by two miners, Alice could cleverly send two transactions where the same amount is sent to both Bob and Adam, and having successfully cheated the system. Let's consider the scenarios where starting from block 1, ten miners are working to create ten different blocks to be attached as block 2. Which block will be attached?

To solve this problem, Bitcoin rules impose that the new cryptographic hash needs to have at least a leading number of zeros, let's say ten for example. The hash has to start with ten zeros or more before the new block can be accepted; otherwise, the miner cannot consider the block for submission.

How can the ten leading zeros be obtained? The miner needs to find an extra input into the SHA 256 machine that will provide the winning output. This new input is the nonce (see Fig. 4.10). There could be more than one nonce that performs the job, but the miners need to find only one.

Let's consider how PoW operates on the example considering the current block 562476. Every miner has the current cryptographic hash for the block

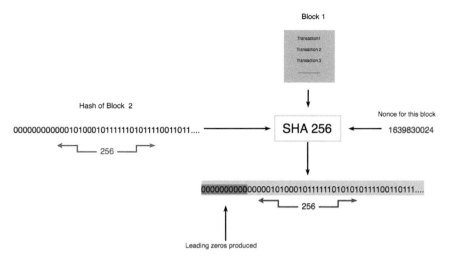

Fig. 4.10 Graphical representation of the role of the nonce in determining the 10 leading zeros in the cryptographic hash

as it was computed as part of the PoW of connecting block 562476–562475. They also have the transactions for the next block they have picked themselves; the third input is the nonce that gets computed by all miners simultaneously, while competing against one another. This process gives the meaning to the word 'PoW' as the miners need to perform a *certain amount of work* to find the nonce. The first miner that possesses a valid nonce will communicate the blocks and the newly found nonce to the rest of Bitcoin nodes.

Now starts the *verification process*; nobody needs to believe the miner, as everyone can verify that the nonce is correct since every node runs a SHA 256 machine. Verification takes seconds. Understanding the difference between creating and verifying a nonce can be compared to writing and reading a book: writing a book can take time, effort and resources, whereas reading one is much simpler.

Miners and nodes do not need to know each other and no human interaction is necessary. In fact, they simply need to understand the rules to play. The Bitcoin protocol automatically checks that the rules are enforced. Interestingly, the PoW gets harder over time. By simply requiring one extra leading zero to the nonce, the miner's work doubles.

As the Bitcoin becomes popular, more miners join, and the probability of getting a correct nonce in a short period of time increases. Therefore, to keep the steady generation of blocks to six per hour, the PoW needs to get harder.

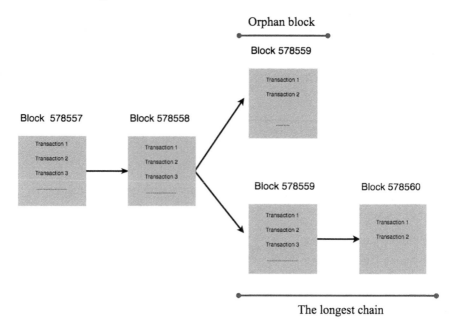

Fig. 4.11 Example of two miners attaching two blocks after block 578558. The likelihood that both branches continue to grow diminishes very quickly. The longest chain survives, while the transactions in the orphan block are reversed

The networks adjust the complexity every two weeks.[11] For our purposes, it's fair to say that PoW is reasonably well designed and does not favour any miner, as it can be shown that the selection of the next block is truly random, like a lottery.

Orphan Blocks and the Longest Chain

Let's return to double spending for a moment. Our friend Eve, being a rather devious character, wants to double spend her bitcoins. She thinks it's possible to send a transaction twice and have it collected by two miners who will both attach a new block—see Fig. 4.11.

If both blocks survive, the attempt to double spend will have succeeded. The rules of the Bitcoin game are that only one of the two nodes will survive. The generation of blocks never stops; the next block will be attached to only one of the two blocks. The likelihood that any two miners will continue to add blocks to both branches rapidly decreases as the ledger grows. Therefore, within a few

---

[11] How this is achieved is beyond the scope of this book. Explaining it in full requires a level of detail more suitable to a university computing graduate.

blocks, the branch with the greatest PoW is determined. The branch of the ledger that survives and remains part of the *mainchain* is called the *longest chain*, and the blocks left behind are called the *orphan blocks*. The transactions in the orphan blocks get reversed, and hence, in Bitcoin, transactions are considered *fully confirmed* only after *six extra blocks* have been attached to the longest chain. Sadly for Eve, this feature makes double spending nearly impossible!

The Bitcoin Network

Miners aren't the only participants in the Bitcoin network. So-called *full nodes* run by volunteers also play an important role in keeping the network operating. These nodes are continuously messaging each other, ensuring that every node has the same information and that local copies of the ledger are updated in real time. Crucially, the nodes validate the block and confirm the transactions. Each node serves on a voluntary basis and the Bitcoin network will continue to operate regardless of whether one node stops, or new nodes are added to the network.

Each node performs identical tasks, and none impacts the operational running of the network as a whole. This is what gives bitcoin its resilience and it is often said to be *censor-resistant*. There is no authority nor organisation that can force any node to run or to stop. Anybody can decide to join the network, provided the correct code is used, and only if all nodes stop running, will

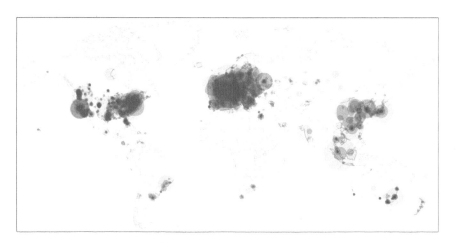

Fig. 4.12 The Bitcoin nodes around the world—The map may not be completely accurate as nodes could hide their activity or anonymise their location (*Source* Bitnodes https://bitnodes.earn.com, accessed February 13, 2019 at 14:15)

Bitcoin cease to exist. Because there is no central authority that determines how a node runs, the bitcoin network is called *decentralised* and most blockchains run on a decentralised or *peer-to-peer network*.[12] A snapshot of the number of nodes running on bitcoin is shown in Fig. 4.12.

The Forks or the Soft Touch of Humans

Bitcoin started with a paper in 2008, and a group of volunteers helped write the code that enabled the first block to be generated. In the last ten years, people and organisations around the world have supported Bitcoin by running a node, or mining, and it is logical to assume that the code has improved in that time. While the technology drives the value and the popularity of Bitcoin, human beings play a crucial part in its development as well.

The community of software developers [28] that works on Bitcoin has a procedure for submitting a Bitcoin Improvement Proposal (BIP). These documents are the means of communicating within the community. The success of the proposal depends on the voting community; a BIP can be rejected, withdrawn or accepted. There are different kinds of BIP[13]; some are developed for information and processes, and others have an impact on the software and, as a consequence, on the interoperability of the Bitcoin.

If a BIP is accepted, developers work to release new software. Nodes will upgrade, and those that don't upgrade will continue to produce an alternative branch. In blockchain jargon, this is known as a *soft fork*. Soft forks, much like a PC or laptop update, will still run old applications. Soft forks do not impact interoperability among the nodes.

By comparison, a *hard fork* introduces new features considered invalid in previous versions. The nodes or miners that will not upgrade will not be able to join the longest chain, hence the ledger will split, and effectively lead to ledgers having two cryptocurrencies. A notable hard fork happened on August 1, 2017, when Bitcoin split into *Bitcoin Cash* (see box). The major difference between Bitcoin and Bitcoin Cash is the size of the block, which separated miners into two groups (see Fig. 4.13).

---

[12]There isn't a single organisation that manages the IT infrastructure, and all nodes in the network have the same rights and privileges.

[13]See for example https://en.bitcoin.it/wiki/Bitcoin_Improvement_Proposals (accessed February 13, 2019).

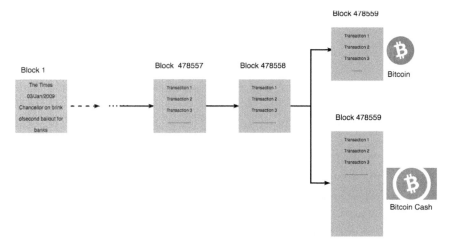

Fig. 4.13 Graph representation of the hard fork done by Bitcoincash

The Birth of Bitcoin Cash

Bitcoin Cash (https://www.bitcoincash.org) is a cryptocurrency created by the hard fork of bitcoin on August 1, 2017. The blockchains of Bitcoin and Bitcoin Cash share the first 478558 blocks; the community split, including miners, and started to generate separate blocks from the 478559th block.

Why did the fork occur? It was driven by humans disagreeing with each other, not by technology.

From December 2017 to early January 2018 demand for Bitcoin spiked and the network was very congested. Block 500008, mined on December 12, 2017 contained 3098 transactions, while block 503000, mined on January 7, 2018, contained 2017 transactions. 2557 is the average number of transactions contained in a block during busy periods, which provides an upper bound of 4.2 transactions per second, which is very slow compared to payment networks like VISA that process over 1000 transactions per second.

To improve the efficiency of Bitcoin, in 2016, the community proposed to double the size of Bitcoin's block, which would double the number of transactions processed. The goal of this change was to to promote Bitcoin as payment network. Part of the community disagreed with the increase of block size, as there were concerns about the time it takes to verify that no double spending has occurred in the new block. As a result the miners separated, and a new cryptocurrency was created.

Querying the Ledger

Because Bitcoin doesn't belong to one single organisation or individual, the ledger must be readable to everyone. There is no confidential information requiring protection in bitcoin, particularly as miners need to know the latest block to keep the ledger growing. As a result, anybody can verify that a PoW is correct, read the ledger and find out how many bitcoins are held at a particular address.

To show how this works, let's take an example of Bitcoin being used to investigate the flow of money by criminals.

In May 2017, the WannaCry ransomware spread across the world impacting several major organisations including the NHS and Nissan (UK), Telefonica (Spain) and FedEx (US). The malware encrypted computer files and only offered to decrypt them after the recipient paid a ransom. Payment was demanded in bitcoins (see Fig. 4.14). We can see clearly that the ransom address

Fig. 4.14 Screenshot of WannaCry ransomware, where the string running on a safe computing environment

is 12t9YDPgwueZ9NyMgw519p7AA8isjr6SMw.[14] While we encourage readers to store their own copy of the Bitcoin ledger on their devices, there are private companies that display the ledger on a website, for example https://blockchair. com/bitcoin; the reader can insert the Bitcoin address into the search bar to find all the transactions associated with this address from the day it was created.

There are companies that obtain information from the ledger to uncover illicit activities such as money laundering. But, you might ask, if Bitcoin is anonymous, how is it possible to link an address to an individual? Bitcoin participants are identified by addresses, not personal details. But, if users publicise their public key, on social media for example, then of course their anonymity is compromised. However, if anybody has carefully protected the public address, is it still possible to link it to an individual or an organisation?

Bitcoin doesn't leak information, but an attacker, with modest effort, could still find ways of linking an address to an individual by:

Spying on the Internet address where transactions take place Transactions leave a digital fingerprint and an astute and competent observer could collate data from various sources and guess who is behind an address. To preserve anonymity on the Internet requires an understanding of the data fingerprint and the techniques needed for anonymity.

Mining the transactions on the ledger Let's assume bad-actor Eve knows that Alice will send 1.25 bitcoin to Bob on April 1, 2019. Eve can check the ledger, and if there is only one transaction for 1.25 bitcoin, she can reasonably assume she has discovered the addresses for Alice and Bob. She could then monitor future transactions and violate Alice's and Bob's anonymity and privacy.

Security in Bitcoin and the 51% Attack

A recent survey [38] suggests that most people believe blockchain is more secure than conventional IT. Are they right to feel that way? It's worth considering what is meant by 'security'. Bitcoin operates on a public ledger: all transactions are public, which implies that there are no secrets to steal. PoW prevents double spending and changes to past published blocks, making ledger transactions tamper-proof. However, 51% of miners could in theory collude to change a ledger's history. To be clear, an attack of this nature couldn't change the rules

---

[14]Three bitcoin addresses are known to have requested the ransom: 12t9YDPgwueZ9NyMgw519 p7AA8isjr6SMw, 13AM4VW2dhxYgXeQepoHkHSQuy6NgaEb94 and 115p7UMMngoj1pMvkp HijcRd-fJNXj6LrLn.

of Bitcoin, or transfer funds on behalf of other participants. It could, however, change the history of the ledger meaning:

1. Some transactions could be deleted.
2. Fraudulent transactions could be added.
3. Double spending would be possible.
4. Some transactions could be ignored and not added to the ledger.

Miners are not the only actors in the network: the rest of the honest nodes could reject the fraudulent blocks with the invalid transactions. The bigger and more diverse the community supporting any cryptocurrency, the harder it is to launch a 51% attack. For social reasons, it is unlikely that diverse organisations will come together to run the attack. At the time of writing, Bitcoin is still the largest cryptocurrency, and the community is sufficiently diverse to make a 51% attack extremely unlikely.

Any attack has the potential to destroy confidence in the blockchain. The community that invests and uses bitcoin (or any other cryptocurrency) is averse to risk and change, as technical risks can hit the value of bitcoin. When the hard fork of Bitcoin Cash was announced in mid-2017, bitcoin's price was hit quite hard. Any suggestions of a 51% attack would likely result in bitcoin losing its value, particularly as it is reliant on the trust of the people who use the digital currency.

Who Generates the Addresses in Bitcoin?

There is no single authority that issues bitcoin addresses. Every user can create their own key pair (public and private key) from which the address is derived. This could look suspicious at first sight, as there could be a risk that two individuals would generate the same keys. In fact, this is (almost) impossible just as it is (almost) impossible to guess a private key by an exhaustive search. The number of keys is so big that the probability of generating two identical keys is so low as to be almost zero. However, random generation can be a weakness in the system and if two users use 'predictable randomness', the statement above no longer applies.

Ten Simple Steps for Alice to Send Bitcoins to Bob

Step 1 The funds: Alice needs a positive balance. Bitcoin does not work on credit.

Step 2 The address: Alice can only identify Bob's address. Any bitcoin sent to the wrong address could be lost.

Step 3 Bitcoin fees: There is no obligation to pay a fee for the transaction. The greater chance of getting the transaction picked the higher the fee. Demand drives miners' fees.

Step 4 Request for the transaction: Alice will send a request to a pool of nodes for a transaction that specifies: Bob's address, the amount to be sent, the value of the miner's fee.

Step 5 Signing the transaction: Alice signs the transaction with her private key.

Step 6 Signature verification: Signed transactions are verified against the sending address.

Step 7 Broadcasting the transaction: The transaction message is sent to the pool of nodes for miners to pick up.

Step 8 Mining the block: Each miner will pick the transactions, which may or may not include Alice's transaction, and will:

- Form and order the transactions in a block and timestamp the block.
- Compute the PoW using the last block and the current block.
- Attach a new block to the ledger, and broadcast the solutions to the other nodes.

Step 9 Waiting for Confirmation: Full nodes confirm the new block is valid by verifying the PoW and ensuring that no transaction is double spent. Sometimes, two or more blocks get attached and only one of the two blocks can remain valid on the main chain.

Step 10 Transaction Confirmed or Transaction Reversed: If after six blocks have been attached, Alice's transaction is fully confirmed and cannot be reversed, and Bob can spend his bitcoin with confidence. Any block attached at the same time will be dismissed and its transactions reversed.

Has Bitcoin Fulfilled Its Original Purpose?

Bitcoin's origin can be traced back to October 2008 when a nine-page paper was published on a public repository by someone called Satoshi Nakamoto [94]. While we will probably never know the author's personal details, email exchanges show they got the Bitcoin network off the ground with the help of a group of volunteers from the Cypherpunk movement [86]. By May 2010, Nakamoto stopped participating in Bitcoin's development and nothing has been heard of him or her since [9].

Further developments have taken place since then, of course, in ways that were probably not originally anticipated. When assessing whether Bitcoin has fulfilled its initial purpose, we need to go back to that nine-page paper from 2008.

Written in a technical style, it used LaTeX instead of other more well-known editors; LaTeX will be very familiar to most scientists and mathematicians working in research and development, or developers. Given the available correspondence, it's reasonable to assume that the author didn't have an academic background. Academics tend to eviscerate a problem from a theoretical point of view before practical development [95]. From the correspondence [9], we know that Bitcoin software development had already started when the paper was published. The bibliography also suggests familiarity with influential work on academic research on cryptographic timestamping techniques [83]. However, no reference is made to e-cash or other seminal works on electronic payment [25, 70, 72]. The paper largely focuses on enabling digital micro-payments from one individual to another without resorting to a financial institution. Nakamoto highlights the trust element implicit in all financial transactions and the fact that systems were getting more expensive while at the same time failing to protect anonymity. It also suggests small transactions no longer make financial sense. It's not clear if Nakamoto refers to cross-border payments or credit card systems, as in the paper only the words 'electronic payments' appear. What is clear is Nakamoto's solution, namely an electronic system based not on trust but on 'cryptographic proof'. Just as with cash, transactions would be atomic, irreversible and anonymous. Nakamoto describes the system where a new transaction is signed with the hash of the previous transaction and with the public address of the owner of the coin put together. It is worth pointing out that the idea of linking transactions via a cryptographic hash was already described in [83]. Nakamoto cited the work and noted a failure to address the double-spending problem, which required a solution. Bitcoin's key innovation, the use of the Hashcash solution [61], prevents double spending and preserves the integrity of the ledger.

A comparison of any recent blog or article about blockchain with Nakamoto's original paper shows how much the language has evolved. The terms 'blockchain', 'blockchain technology' or 'Distributed Ledger Technology' do not appear in the original paper, which doesn't make any reference to the financial collapse which happened just a few months earlier.

When Bitcoin was first created, there were no miners other than Nakamoto, and Bitcoin had limited value as a currency. The blockchain was, in fact, insecure because of the absence of mining and anybody could have easily overwhelmed the process [95]. Ten years later, Bitcoin has a market capitalisation of over $210 billion, a thriving community and an ecosystem that supports the currency.

Despite this, it's legitimate to ask: has it fulfilled its original purpose? While the software has been upgraded by the community, in its key point it is still what Nakamoto envisaged in the article. The economics of Bitcoin have taken a route that would have been very difficult to anticipate in 2008. Mining bitcoins is no longer open to everyone: to run a profitable mining company today requires a serious investment in hardware. Miners have buildings full of specialised processors that can rapidly compute a nonce [99]. It is estimated [8] that the amount of energy currently consumed by miners in one year is equivalent to the power uptake of a small country such as the Czech Republic!

It is also estimated that the average cost of mining a bitcoin is around $5000, which gives an idea of how the price of bitcoin is likely to evolve. Keeping the price of bitcoin below $5000 will remove some of the financial incentives for miners. If miners were to turn their backs on the system, it wouldn't spell the end for bitcoin, but it would mean fewer enterprises and less hardware-intensive mining activities. More comprehensive studies also show that miners are using renewable energy sources [99].

Regardless of how the price of bitcoin will evolve, the cost of mining each bitcoin has to be lower than the price of bitcoin that the market is ready to pay. Today, miners are located in areas where electricity is cheap, and average temperatures are not too high, helping to prevent processors from over-heating. Miners tend to hide their location, and a recent study [99] suggests that facilities could be located in China, North America (USA and Canada), and North-Eastern Europe (Russia and Georgia). Chinese miners have been responsible for the majority of mining, and it seems that this trend is not going to change soon (Fig. 4.15).

Bitcoin fees are typically a few cents (of a US dollar) per transactions; however, between December 2017 and January 2018, when the bitcoin price was close to new heights and the network was congested, fees exceeded $30.

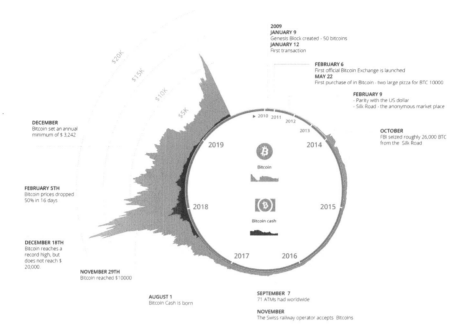

Fig. 4.15 Graph of the Bitcoin and BitcoinCash price, and historical events from January 1, 2014 until June 19, 2019 (*Data source* https://www.cryptocompare.com/)

A morning coffee purchased with bitcoin would have attracted an exorbitant fee to enable the transaction to be prioritised.

In reality, Bitcoin has become an economic game between two distinct participants: users and miners. Users need miners to maintain a ledger and to enable bitcoin to remain a decentralised payment method. Miners need users to fulfil the bitcoin purpose. The rules behind this economic game, however, seem to be creating unintended consequences: as bitcoin becomes more popular, the price increases, but because there is an upper limit to the number of transactions that can be processed, fees rise exorbitantly so that in the future it would be impossible to justify using bitcoin for micro-payments.

Summary

Blockchain technology was born with Bitcoin and Bitcoin is a form of blockchain. While blockchain and cryptocurrencies come in all shapes and forms, there are three elements common to all blockchains.

- The transactions: moving assets/coins from one participant to another
- The consensus rules that determine how the ledger can be updated, PoW in the case of Bitcoin
- Ledger or a secure chain of blocks with an immutable history of all transactions.

Name	Value
Bitcoin	1 BTC
	1/10 BTC
bitcent	1/100 BTC
millibit	1/1000 BTC
	1/10000 BTC
bit	1/100000 BTC
finney	1/1000000 BTC
satoshi	1/10000000 BTC

Transfers of bitcoins from one user to another are signed, using asymmetric key cryptography, and recorded in the ledger. The transactions' history on the ledger is secured by the PoW, in a way that attempts to change a past transaction will also require a recalculation of all subsequent blocks. Cryptocurrencies have built-in security measures that prevent double spending, and unlike fiat currencies, rules are enforced via the use of technology and, more importantly, without relying on a third party.

The supply of bitcoin is managed by an algorithm according to the following rules:

1. The total supply of bitcoin is fixed at 21,000,000.
2. New bitcoins are produced on average every ten minutes, and this will continue until the full number of 21,000,000 is achieved.
3. Every successful minted block delivers a newly minted number of bitcoins called the reward.
4. After 210,000 new blocks are produced, the miners' reward is halved.

Blockchain offers a number of benefits including transparency and a single view of the truth.

Frequently Asked Questions

Has bitcoin any value?
Bitcoin's value is in its usefulness as a method of payment within a community.

How is the price of bitcoin determined?
Bitcoin is not run by a corporation and it is not backed by any asset or service, and its price is computed in a similar way as other currencies, by considering supply and demand in the markets.

A study from the Bank of England [66] identifies the following criteria to understand the demand of bitcoin:

1. The expected real return on bitcoin
2. The benefits of using bitcoin as a medium of exchange to purchase goods and services
3. Ideological preferences
4. Specific view on the popularity of Bitcoin.

Positive or negative news influences people's expectation, and announcement of technical changes in the code can also have an impact on the price, just as announcements by the Federal Reserve have an impact on the value of the US dollar.

Can I buy a fraction of bitcoin?
Yes, bitcoin has eight denominators as shown in the table to the right.

What are the risks in buying bitcoin?
Buyers need to be informed. It's important to keep in mind that it is relatively new technology, and while it appears secure there are associated risks. Like cash, Bitcoin transactions are irreversible, so if you send Bitcoin to the wrong person it's impossible to get it back—unless the recipient is willing to return it. Bitcoin's price is determined by the number of participants that accept it but there are no guarantees that will continue in the long term. Because it's a relatively new currency, it's difficult to forecast a future stable price. If bitcoin's price goes down too much, there will be a loss of trust and mining might not be sustainable. If the price goes up too much, it could become prohibitive to use. Anybody entering the market needs to be aware that the bitcoin price fluctuations are very high. See Fig. 2.4, which shows the volatility of the daily change in Bitcoin price.

How do I get hold of bitcoins (or other cryptocurrencies)?
There are several ways:

1. You can mine bitcoins.
2. You can ask somebody to send you some bitcoin (in exchange for fiat currency or other services).
3. You can get bitcoin at an ATM machine[15]

---

[15]Consult https://coinatmradar.com to find out if there are ATM close to you (accessed September 11, 2018).

4. You can use an online exchange that will allow you to buy bitcoin using your credit/debit card.

What happens if I lose my private key?

You won't be allowed access to your Bitcoin account to transfer or exchange bitcoins, unless somebody can either find the private key of an account, or miraculously guess the private key, the account is locked forever.

How do I open a Bitcoin account?

Bitcoin is not a bank or a financial institution where you can open an account to keep your cryptocurrency. You simply create a valid public address and start working with that.

What is a Bitcoin wallet?

A wallet is an application with a user-friendly interface that allows you to receive, store and send bitcoins.

What can my organisation do with bitcoin?

Your organisation can:

1. Accept payment in bitcoin.
2. Use the Bitcoin network as an alternative payment network and transfer bitcoin to your customers or employees.
3. Use it to timestamp documents.

Do I need to be a miner to send bitcoin?

No. The miner's goal is to attach a node to the ledger and collect fees. You are encouraged to run a full Bitcoin node. You can use your laptop to request a transfer of money via a piece of computer code called a wallet. There are several websites that have a user-friendly interface that will enable you to send a transaction to a node.

Do I need to run a full node to send bitcoin?

No. You can use the wallet provided by third parties; however, running a full node provides extra security as you will have a copy of the full ledger and you can be certain of operating according to bitcoin rules.

Do I need a wallet to send and receive bitcoin?

No, but the wallet makes that easier.

I've heard scalability is a challenge for Bitcoin. What does this mean?

Bitcoin can only process about 4.2 transactions per second whereas Visa can process more than 1500 per second. For mass adoption, bitcoin needs to increase the number of transactions it can process.

Could the PoW be replaced by other protocols?

The PoW is not eco-friendly as it consumes a lot of energy; new protocols are being devised that will provide a similar lever of security but consume less energy. A strong contender is *Proof of Stake* (PoS) (see Chapter 5). In PoS, miners effectively collateralise their participation with stakes, which needs to be forfeited if they act maliciously.

Do all cryptocurrencies use the PoW?

The PoW is a mechanism that ensures consensus, thereby verifying the legitimacy of the new block attached to the ledger. Different blockchains running on different networks can use other consensus algorithms.

Why is it important for bitcoin to be a peer-to-peer network?

Because you can trust the value associated with the coins, as it is not possible for one single organisation to prevent Bitcoin from operating. The whole community, which is currently is located all over the world, needs to disappear before bitcoin will stop operating. Bitcoin is censor-resistant for the same reasons.

5

The Rise and Rise of Cryptocurrencies

At the time of writing, there are a few thousand cryptocurrencies[1] (or virtual currency) in existence, which is a staggering number given that the first cryptocurrency cryptocurrency was only created in 2009. It would be a never-ending and pointless task to try and analyse each cryptocurrency, as some are added every day. Instead, we take a practical approach and provide a broad categorisation of how they are created and what are they used for.

You will probably have questions. Two that are most frequently asked are: Is it easy to make money buying and selling cryptocurrency? And: Will it ever become commonplace to pay for goods and services with cryptocurrencies? In this chapter, we will try to answer these questions and shed more light on how cryptocurrencies can be used for some business.

How Do I Use Bitcoin?

Created in the wake of the global financial crisis, Bitcoin was the first-ever cryptocurrency. Like others that followed, it can be used in a variety of ways, ranging from the mundane, such as buying a loaf of bread, to acting as the currency of choice for migrant workers wanting to transfer money home to their families. It is popular with speculators, who see an opportunity to make a lot of money trading in cryptocurrencies. In fact, the latter is by far the most

---

[1]Different data sources provide conflicting information. For example CoinMarketCap https://coinmarketcap.com claims over two thousand while CoinGecko (https://www.coingecko.com) claims over five thousand.

© The Author(s) 2020
M. G. Vigliotti and H. Jones, *The Executive Guide to Blockchain*,
https://doi.org/10.1007/978-3-030-21107-3_5

common use of cryptocurrencies, with current research showing only 11% of it used for payments [84].

Use of cryptocurrencies has at least tripled between 2016 and 2018,[2] mainly by individuals; hedge funds and merchants account for the majority of business users. US$-bitcoin trade accounts for approximately half of all volumes, followed by the Japanese Yen (21%) and the Korean Won (16%). Long-term investors are mixed and include high-net-worth individuals, new crypto-focused investment funds and more traditional funds [84]. Bitcoin futures have become an integral part of the cryptoasset investment landscape since Chicago-based exchanges CME Group and Cboe[3] started offering cash-settled bitcoin contracts in December 2017. These products allow industry actors, such as miners and payment service providers as well as investors, to hedge the volatility that is inherent in cryptoasset markets, while helping to bolster stability and maturity.

Returns on bitcoin and other cryptocurrencies are spectacular, particularly when set against the backdrop of the deepest and most prolonged financial slump since the Second World War, and record low interest rates both in the USA and Europe.

The facts behind the headlines are difficult to exaggerate. A relatively modest investment of $1200 in bitcoin in January 2013 would have seen a return of $1,806,000[4] in December 2017, a staggering 150500% increase. However, it's worth injecting a note of caution at this point because it's also true that someone buying $1200 worth of bitcoin in December 2017 would have seen the value plunge to less than $200 in February 2019. If nothing else, these figures highlight the volatility of this still comparatively immature currency. What goes up can come down—occasionally with a crashing thud.

Putting aside its investment potential, it's worth asking why anyone would want to use bitcoin as a payment method, particularly at a time when paying for things using conventional currencies can be done with a smartphone app or through a click of a mouse on a laptop.

Convenience is one answer. Within the European Union,[5] no-fee electronic money transfers take about two hours. If you think that's fast, in the UK transfers can be made within seconds,[6] thanks to the introduction of the Faster Payment Service (FPS) in 2008. Transfers from the UK to EU countries that use the Euro take a little longer because the UK is not part of the Eurozone and

---

[2]This is a lower bound as we considered only users verified by crypto-financial institutions.
[3]Discontinued in March 20, 2019.
[4]Source https://coinmarketcap.com.
[5]Strictly speaking within the Single Euro Payment Area (SEPA).
[6]This is true for the participating banks and building societies, which at the time of writing, represent 95% of the total payments in the UK.

money transfers attract additional exchange rate fees. However, the situation becomes much more complicated when moving money further afield. It can take up to one week to complete a transfer to some countries and on top of that the sender can be hit with eye-watering fees of more than 10%[7] of the amount transferred. In some cases, particularly in the developing world, a poor banking infrastructure—or even none at all—makes it difficult if not impossible to transfer money across borders. In these circumstances, it's hardly surprising that bitcoin, which can be transferred relatively inexpensively within an hour, is seen as an attractive alternative—particularly when an Internet connection is all that is needed to complete a transaction.

Even in countries that have good banking infrastructure, some people still use bitcoin for ideological as well as practical reasons. In countries with questionable civil rights, where it might be commonplace for governments to spy on their citizens, Bitcoin offers a welcome degree of anonymity for transfers that bypass national payment systems. In Venezuela or Zimbabwe, for example, where a combination of political instability and spiralling inflation has had a devastating impact on the economies of both countries, Bitcoin is used as a stable means of exchange, instead of the increasingly worthless national currency.

More Than One Way to Buy Your Daily Bread—Or Chocolates!

To see how this works in practice, let's say hello again to our friends Bob and Alice, who we met in previous chapters.

As shown in Fig. 5.1, Alice, a cryptocurrency novice, wants to send a box of fancy chocolates to her friend Bob. Unfortunately for her, the artisan chocolatier she has chosen only accepts bitcoin. Undeterred, Alice pushes on with her purchase but needs to understand how to pay for the chocolates.

Her challenge, as she sees it, is getting hold of bitcoins to pay the chocolatier for his splendid chocolates. After some thought, Alice takes the bold decision to run her own bitcoin node (a node is a computer or server running Bitcoin software), making her part of the peer-to-peer network. As a member of the network, her node will check that transactions are in the correct format and forward information to other nodes to ensure the ledger is synchronised.

---

[7]The fees for remittances.

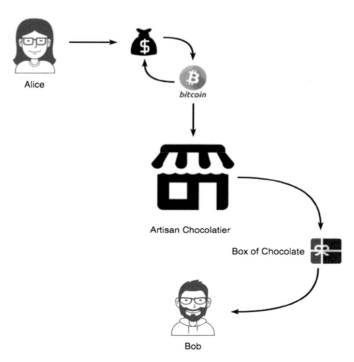

Fig. 5.1 Buying chocolates with Bitcoin

She will also be able to submit her own transactions. However, running a node will not automatically provide her with bitcoin and she will need to take extra steps to insert bitcoin into her newly created public address.

Alice's options are:

Mining This requires specific software. If Alice wants to keep her mining business relatively small, her chance of attaching a block will be on a par with buying a winning lottery ticket. They would improve if she joined a pool (making her computational resources available to other miners), but just as lottery syndicate members have to share their winnings, Alice would receive bitcoin proportional to the quantity of work her processing power added to the pool. She would also have to wait for hundreds of confirmations, or blocks subsequently to be added before being able to spend the newly minted bitcoin.

Community Exchange If Alice decides mining isn't for her, she will have to look for someone who already has bitcoins they are willing to exchange. In our example, Eve has bitcoins and exchanges them for Alice's dollars.

Fig. 5.2   A Bitcoin cash machine in South London (Photographed in July 2019)

In many ways, this is just like a normal exchange of cash: Alice gives Eve her Bitcoin address, waits for the bitcoin to arrive, and then gives Eve her dollars in exchange. Alice will have to pay the miners' fee and make a judgement on a fair price for the exchange (there are a number of online resources to help her do this), but ultimately it all depends on how much an individual is prepared to pay.

Cash Machine Alice's third option is a cash machine where she can insert cash, or a bank card and bitcoins will be sent to her public address. It's a familiar transaction for globetrotting Alice who often uses her bank card to get foreign currency on overseas trips. In this instance, she would pay a miners' fee and a fee at the cash machine.

Assuming Alice has chosen one of these options, she's ready to use her node to complete the transaction with the chocolatier (see Chapter 4, Table 1). Alice decides what option is best for her based on her views on Bitcoin.

She may believe in it in principle; support decentralisation (as opposed to conventional banking); wish to remain anonymous; or want full control of her bitcoins. This last reason might be particularly relevant if trust in financial institutions has collapsed, due to war or economic decline. Under these circumstances, having an alternative method for moving money across borders becomes crucial.

We know that Alice is dipping her toe into the Bitcoin water because it's the only way she can pay for Bob's artisan chocolates; because of this, she might

Type of Service		Description	Features	
Brokerage Service	⇆	Enable users to buy and sell cryptocurrencies at a given price and makes profit by the spread between the buying and selling price.	Comply with AML legislation - customer identity is verified resulting in loss of anonymity for users Trade crypto-to-fiat or fiat-to-crypto Credit balances (for crypto and fiat) are kept on behalf of customers. Exchange has custody of crypto assets and associated private keys	Centralised
Order-Book Exchanges	🏛	Matches buyers and sellers via an automatic trading engine. Charges a fee for matching service.		
Advanced Trading Services	📊	Allows users to buy portfolio of cryptocurrencies - attractive for speculators.		
Peer-to-Peer exchanges	💻	Connects users via a web-site. The way in which it operates varies.	Operating on an ad-hoc basis on a web-site as people freely join and leave.	
'Decentralised' Exchanges (DEX)	⚛	Trading of crypto assets is performed via a blockchain. It:	• Leaves the custody of the cryptocurrency to the users • Uses a blockchain for order matching among users • Uses a blockchain (similar to bitcoin) for settlement	Decentralised

Fig. 5.3 Overview of the cryptocurrency ecosystem [84]

want to use what is known as a cryptocurrency exchange. Exchanges come in many forms. For simplicity, Alice can opt for a fully-fledged enterprise exchange that takes some fiat currency—such as dollars or pounds—and delivers bitcoin at a pre-agreed rate. Exchanges of this kind are subject to money-laundering regulations (see Chapter 9), meaning Alice would have to reveal her identity and other personal information. On the plus side, however, if the exchange operates in a country with a stronger consumers' law, and it is registered, the exchange has a contractual obligation to look after her account, and in our case make sure the chocolatier receives his bitcoins. Alice could opt for this method because it seems no more complicated or riskier than Internet banking.

From this, we can see that fiat currencies can be exchanged for bitcoin with relative ease. Choosing an exchange can be more of a problem, however, given the choice that is available (see Fig. 5.3).

> ### Should You Trust a Centralised Exchange?
>
> According to research, hackers stole $1.5 million from exchanges between 2011 and 2018 [84]. How it is that possible?
>
> Unregulated crypto-exchanges don't have to follow best practice. The most infamous case is Mt.Gox, which was one of the biggest cryptocurrency exchanges in the world, until it closed in 2014 after a series of cybersecurity attacks on the private keys of its Bitcoin account holders. Opened in 2010, Mt.Gox—which stands for: Magic: The Gathering Online eXchange—expanded rapidly. By 2013, it traded 80% of all bitcoin in circulation. Yet by February 2014, Mt.Gox filed for bankruptcy protection in Japan (and later in the USA). Some 744,408 BTC belonging to customers and a further 100,000 BTC held on the exchange had been stolen. Mark Karpelés, who bought Mt.Gox in 2011, was charged with fraud and embezzlement and investigations are ongoing, even though the hacker has been identified and arrested.
>
> Mt.Gox's security was so lax that 200,000 private keys, initially considered stolen, were later found almost by accident. The investigation revealed that the private keys were unknowingly hacked.
>
> This extreme example shows that unless companies take cybersecurity seriously, private keys will be hacked.

Although the use of cryptocurrency is still relatively new, it is entirely possible to imagine a world where they are so widely used that exchanges aren't needed. The price of goods and services would be determined by availability, as is presently the case for agricultural products. For example, in Italy tomatoes are easy to grow and are relatively inexpensive, while salmon is not common and is expensive. In Norway, of course, this is reversed; this decides their local prices.

Because there is no economy associated with bitcoin or other cryptocurrencies, we cannot use bitcoin to compare goods and services. The value in bitcoin of goods and services is merely the exchange from the native currency into bitcoin, where the exchange rate is decided by supply and demand.

Cryptomarket—An Overview

Most of the two thousand-plus cryptocurrencies in existence today use blockchain or a variation of the technology. Exponential growth started in late 2015, and new cryptocoins are added every day. How it is that possible?

Understanding how cryptocoins are created is key to understand the market. The four common ways to create cryptocurrencies are:

Reuse of code Most cryptocurrency and blockchain projects are open source with licences to easily use or modify code. This makes it easy for anyone to take, for example, the Bitcoin software and create a new cryptocurrency, a so-called *altcoin*. Examples of cryptocurrencies created in this way are Litecoin, Zcash and dash which are *private coins*, meaning there is an extra level of privacy for the users (see Appendix A "Useful Resources"). For any project to be successful, the biggest challenge is to create interest from users and attract miners willing to mint new coins.

Creating a blockchain With the right skills and adequate time and resources, a new cryptocurrency can be created from scratch. Yes, it's time-consuming but it has the advantage of improving on what already exists. Bitcoin, for example, has a number of limitations including:

Scalability It can only handle a maximum of seven transactions a minute.

Lack of anonymity for users It's relatively easy to identify someone by simply monitoring the ledger.

Limited use While Bitcoin can be used as a payment system or for small data storage, no extra functionality is possible.

Cardano, Tron and Ethereum are examples of cryptocurrencies created to improve both functionality and scalability: users are allowed to run generic software applications on the blockchain; Monero is a private coin.

Creating a token or new cryptocoin from an existing platform More advanced platforms like Ethereum enable people to create, in an easy way, new software applications that run on the blockchain. A very popular way is to write an application that generates new coins, often called *tokens*. Tokens are exchanged on the blockchain infrastructure where they have been created, meaning that, differently from altcoins, they deploy the miners from the *native blockchain*.

What does this mean? Let's consider Alice and Bob for a moment: if Alice generates a token on the Ethereum blockchain—let's call it AliceCoin—and she wants to send 10 AliceCoins to Bob, the Ethereum miners will mine Alice's transactions together with all the other Ethereum transactions. Alice will pay a fee to record the transfer into Bob's account.

A new phenomenon, known as Initial Coin Offerings (ICOs), is essentially a crowdfunding activity where companies exchange their own tokens in exchange of funding (see Chapters 8 and 9). As discussed earlier, one such platform is the Ethereum blockchain. It is fair to say that 90% of tokens are created in this way. The most common tokens are created under a standard called *ERC20*, which ensures interoperability with the rest of the Ethereum ecosystem; other standards are used but aren't as popular.

Hard Fork Starting from a block, two independent ledgers are created. This is the case of Bitcoin Cash (see Chapter 4, section 'The Forks or the Soft Touch of Humans') and Ethereum Classic—(see Box 'The Darkest Day in Ethereum History').

The Darkest Day in Ethereum's History

Block number 1920000, mined on July 20, 2016, changed the fate of the Ethereum blockchain. Two separated cryptocurrencies started to develop: Ethereum and Ethereum Classic. 'The DAO' hack, as it became known, was one of the darkest days in Ethereum's short history caused the separation. The Decentralised Autonomous Organisation (DAO) was created as a token by a company called Slock.it, and sought community funding. The DAO token provided holders with voting rights on how to spend funds. This was seen as a significant opportunity to implement democracy through technology. The DAO raised an astonishing $150m from more than 11,000 enthusiastic members—far more money than its creators expected.

However, the application used for The DAO a smart contract (see Chapter 8) wasn't secure and by June 17, one unauthorised user, someone who had not participated in the initial funds, managed to drain more than $3.6 million Ether into what is known as the 'child DAO'. The user had to wait 28 days before spending the funds, so the community had time to put remedies in place. The majority decided to take action to recover the money, while the others decided to use the original ledger, which became the Ethereum Classic. Ironically, the Ethereum Classic is now the less valuable blockchain, despite remaining faithful to its original aim.

Blockchain or the code is not law, as crypto-enthusiasts claim, and governance of the blockchain determines the future of these platforms.[a] The developers of Ethereum wrote a new smart contract to enable legitimate token holders to get their funds back. Ethereum Classic shows how ultimately any decision to change a code is down to the participants involved, in other words: it depends on people, not software.

---

[a]Ethereum and Ethereum Classic's explorer block 1920000 is identical (see https://etherscan.io and https://gastracker.io) but from block 1920001 the two blockchains differ.

Fig. 5.4 Graph of the Ethereum and Ethereum Classic price, and historical events until June 19, 2019 (*Data source* https://www.cryptocompare.com/)

A significant evolution from blockchain is *stablecoins*, whose value is linked to commodities such as oil, gold, diamonds or a conventional fiat currency. Their purchasing power is stable (hence their name), in marked contrast to other more volatile cryptocurrencies. Stablecoins are mostly used by traders or small exchanges that can't legally hold fiat currencies. They are also deployed by financial institutions for settlements, and they can help with adoption for payments as it reduces the risk for a merchant. In the latter context is placed *Libra* [54], recently announced by Facebook which will be launched in 2020 and will be pegged to a basket of cryptocurrencies (see Chapter 10). A troubled stablecoin is Tether, loosely linked to the US dollar with the price rarely straying from parity, due, so Tether claims, to one dollar being kept in a bank account for every Tether issued. This has been questioned, and in 2018, the US Security Exchange Commission launched an investigation to ensure Tether's backers held the dollar reserves they were claiming.

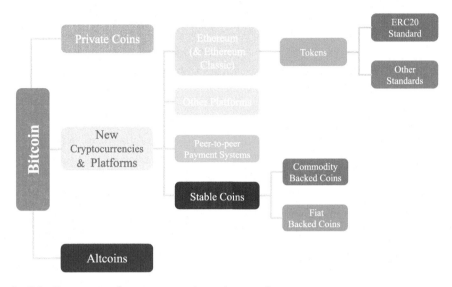

Fig. 5.5   Taxonomy of cryptocurrencies, tokens and payment systems

There are also the *peer-to-peer payment systems* like Ripple 'A.2', where participants are typically financial institutions that get connected, via a number of hops, to other trustworthy parties. In other words, Ripple connects to institutions that do not know each other through a set of middle parties who have pre-existing relationships.

The graph in Fig. 5.5 shows the taxonomy of the cryptocurrencies we have already considered; it only describes cryptocurrencies built on public and permissionless blockchains; and it does not consider enterprise blockchain.

Cryptocurrencies, tokens or cryptoassets differ from each other in price, number of transactions per day, the size of the community supporting the project and, crucially for the security of the project, the number of miners. Market capitalisation provides a good indicator of the popularity and likely longevity of a cryptocurrency. The graph below shows the market capitalisations of the twelve most popular coins. Not surprisingly, Bitcoin is the most capitalised, meaning that investors and users are prepared to inject the greatest amount of money into this, the first-ever cryptocurrency. Details of the cryptoassets discussed here can be found in section Appendix A "Useful Resources".

> ### Scalable Consensus—Proof of Stake
>
> It is virtually impossible to list all features in new blockchain platforms, but it is worth considering the evolution of the consensus process as it is a distinguished feature.
>
> Proof of Work (PoW), the consensus process deployed by Bitcoin, is costly, time-consuming and uses a great deal of energy; combined, these factors limit the scalability of the blockchain (see Chapter 4). Many of the newer cryptocurrencies have been set up to take on the challenge to overcome these limitations and have devised a new process to enable participants to agree on an updated version of the ledger.
>
> One such consensus is *Proof of Stake* (PoS), a type of algorithm, which instead of let miners compete against each other, selects the next node that will attach a block to the ledger. How does the node's selection happen?
>
> A node willing to participate in the selection process is required to lock a certain amount of cryptocoins into the network as its *stake*. The size of the stake determines the probability for a node to be selected as the next validator: the higher the stake, the higher the probability of becoming the next validator. The job of the selected validator is to check the validity of the transactions in the blocksign and attach the block to the ledger. The stake, which was kept locked during the validation process, is only released if independent nodes can verify that no double spending occurred. As a reward, the chosen receives the transaction fees. There are several versions and modifications of the PoS that fundamentally operate on the principles described here.

Maintaining Privacy

How can you use cryptocurrency and protect your privacy? Whatever cryptocurrency you are sending, be it monero or bitcoin, your experience will be much the same. First, you insert the address of the recipient into the wallet, stating the number of coins you want to transfer and the miners' fee you are willing to pay. Both ledgers are public, and anybody can read them. How is it possible that Monero is more private than Bitcoin? Both of them do not require people to disclose identity.

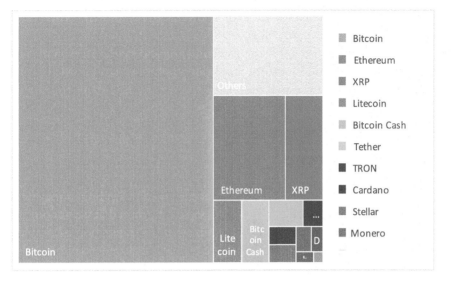

Fig. 5.6 Market capitalisation of the cryptocurrencies discussed in this chapter. Bitcoin dominated the market followed by Ethereum and Ripple (*Data Source* Cryptocompare https://www.cryptocompare.com/)

On the Bitcoin ledger, the information that is visible for a transaction is

1. The originating Bitcoin addresses
2. The recipient's Bitcoin address
3. The amount of bitcoin transferred
4. The fee paid.

The Monero ledger, by contrast, reveals very little, only the miners' fee. Addresses and amounts sent are excluded. So there are significant differences between private and anonymous coins. With the latter, the owner's identity isn't revealed but transaction details are; a private coin conceals all transaction details. Without going into detail, there is some clever cryptography behind private coins that would enable the sender to prove who initiated the transaction, without the rest of the world finding out—unless of course, the parties want to disclose it.

Looking back to 2012 and 2013, Bitcoin was used to trade drugs and other illegal goods on a website called Silk Road, damaging Bitcoin's reputation in the process. There's no hiding from the fact that a private coin is an attractive target for this type of illegal activity and money laundering. It reignites the debate about regulation: a person's right to privacy and the desire to keep cryptocurrency transactions anonymous doesn't make them a criminal; but, nevertheless, some criminals will exploit this. There is no easy answer to the conundrum and the debate will go on.

Corporates will be interested in protecting privacy as well. An example is the J.P. Morgan Blockchain Quorum, that is incorporating zero-knowledge proof technology used in the Zcash [14] to empower the Interbank Information Network (IINSM) to enable fast cross-border payments.

Payment Platforms

In the last twenty years, we have become accustomed to sending money with a click of a mouse or a tap on a smartphone screen. Beyond that we put our trust in banks (and sometimes pay a fee) to ensure that the transaction is completed quickly and efficiently. Is there any chance that crytpocurrencies or blockchain technology could make even a small dent in this tried and tested method? We will take a look behind the scenes to see how banks operate, so we can appreciate the risk and potential benefits of blockchain operation.

As a starting point, let's consider two banks—one wants to send money and the other wants to receive it. If this was done through a cash transaction, a clerk from the sending bank would deposit the money with a clerk at the receiving bank. But, while this may seem the simplest of transactions, where is the proof that the exchange happened? And in another scenario, what if the sending bank has a last-minute change of plan, leaving the receiving bank short of cash? To ensure fairness, a simple process could be devised where:

1. Before meeting, the two clerks agree on the amount exchanged
2. The receiving clerk checks the notes are not counterfeit
3. The clerks together count the money to ensure the correct amount of cash is exchanged
4. Both clerks sign a receipt
5. Each clerk signs to the other a piece of paper confirming the time and the location of the exchange.

A payment system is a set of rules and procedures, which can be performed electronically or manually. Regardless of how they carry out, a payment system should be *fair*, no party should take advantage from the other. In the above example, no bank shouldn't be able to repudiate the agreement, or be able to reverse the transfer of money, after completion.

Current Payment Systems

There are several bank payment systems in the world and how they broadly work is set out below. Bank accounts are essentially accountancy tools that link individuals or institutions to a deposit or a debit, which contributes to a bank's liabilities and assets. Transfers between banks increase and decrease liabilities.

Transferring Money in the Same Country

If Alice and Bob both have bank accounts at BankForCryptographers, and Alice wants to transfer £10 to Bob, then the bank will reduce Alice's account by £10 and increasing Bob's account by the same amount. From the bank's perspective, its liabilities and assets remain the same, it's just an internal change in the internal bank's ledger.

If Alice moves her account to BankForMathematicians and wants to transfer £10 to Bob who has remained loyal to BankForCryptographers, then the situation is more complicated. A simple solution is possible because BankForMathematicians just happens to have an account with BankForCryptographers. This means Alice can send money to an intermediate account at her new bank and Bob eventually receives £10 via this account, in the following way:

- BankForMathematicians reduces Alice's balance by £10.
- BankForMathematicians adds £10 to the account held by BankForCryptographer.
- BankForMathematicians tells BankForCryptographer that its account has been increased by £10 and instructs it to increase Bob's balance by £10.

Without going too much into detail, the movement of money between BankForCryptographers and BankForMathematicians happens through domestic payments mechanisms. In the UK, for example these are Faster Payment Service (FPS), Chaps, Bacs [2], *Real Time Gross Settlement* (RTGS) which are typically settled with the Bank of England,[8] The key point for our discussion is that, in traditional payments, each bank keeps its own ledger. A customer requests a payment, and the banks inform each other of the intentions to make changes. Funds are withdrawn from the sender's account to prevent double spending; however, final settlement happens with the help of a neutral third party that minimises the risk, and in the UK is the Bank of England.

---

[8]In the UK, the RTGS and Chaps are run by the Bank of England.

Cross-Border Payments

We have already seen how Alice has transferred money to Bob between accounts at the same bank and then between different banks in the same country. Now she wants to send him foreign currency, $10, from a US account. Alice's bank needs an account with a bank that can clear and hold US dollar for it; the US dollar correspondent clears with Bob's bank on behalf of Alice's bank.

The transfer would happen roughly as described above with a few complications:

1. A conversion rate has to be agreed before updating the relative accounts in the UK and US banks, and
2. There's a transfer fee to support the operation in another country.

The main difference is that there is no central bank supporting clearing and settlement. There are, however, clearing and settlement institutions that operate in pairs. Where such clearing and settlement facility does not exist, the banks have to agree on changing their position at the same time. As a consequence, transactions can only occur during banking hours and counterparty risks can be a big hurdle.

Why is this more challenging? Firstly, counterparty risk is an obstacle in establishing inter-bank relationships: the higher the counterparty risk, the less likely it is that other banks will engage in trading relationships. It also increases transfer costs. This explains why cross-border payments in developing countries are particularly expensive.

Secondly, where there is no direct relationship between originating and receiving banks, a number of intermediary banks are required to build a connection. For example, for each transaction into a foreign currency requested by a customer at the BankForMathematicians, the bank must have a corresponding account in the destination country. If Alice wishes to transfer £100 into Ghana Cedi, then the BankForMathematicians must have a corresponding banking relationship with a bank in Ghana. As there are 180 countries in the world, most major banks have accounts in major currency countries, such as the US, UK, EU nations, Japan and Switzerland. If, however, someone wants to send money from Angola to Turkey, and there is no relationship between the BankofAngolianMaths and the BankofTurkishMaths, the former will have to find a bank that will enable a connection. For example, BankofAngolian-Maths could have a relationship with a bank in South Africa, which in turn could have a relationship with a bank in the UK, who has a relationship with BankofTurkishMaths.

Bridging currencies relies on the economics of demand for the exchange because of relationships locally, so the transfer will go from Angola to Turkey via South Africa and the UK. Why? Well, Angola will have trade with South Africa—so there is liquidity—the national currency of South Africa is relatively liquid against the US dollar, which in turn is very liquid with the Euros, the latter being liquid with the Turkish Lira. Phew! More to the point, in every exchange, the cost of the transfer increases because:

1. In buying and selling various currencies the customer bears the cost of the spread
2. Each transfer bears a fee
3. It takes time to settle each trade.

This example shows that with each intermediary, the cost of the transfer goes up and it takes longer to complete. The World Bank estimates that cross-border payments performed by banks carry an average fee of over 10% of the sum exchanged—considered from 2008 until Q2 2019 [31]. For low-paid migrant workers who regularly send money home in small amounts, this represents an enormous expense. Money transfer specialists such as Western Union or Ria offer an alternative, which can be faster than banks, although still expensive with average remittances of almost 7.5% in 2018.[9] Remittances are a fundamental part of the economy in emerging economies, contributing as much as 10.5% of GDP in the Philippines, and almost 30% in Haiti. Global remittances reached a record $642 billion in 2018. Two years earlier, the official development assistance globally reached $158 billion.[10] A reduction of costs in cross-border payments by just 5% will result in annual savings of $16 billion [31].

Payment Through Blockchain

Payment with blockchain would not follow these models of operation. First of all, the need for a third party for settlement (and clearing) is due to the fact that each bank keeps its own ledger and there is a gap between the moment the payment is agreed and the moment the payment is actually settled. The third party has the role to prevent counterparty risks.

With a cryptocurrency, there is only one ledger shared among all participants. There is no difference between the agreement of a payment and the settlement: they are the same thing, so there's simply no need for a third party

---

[9] In some countries such as Japan or South Africa the average remittance cost is well above 10%.

[10] See https://data.worldbank.org/indicator/DT.ODA.ALLD.CD.

for settlement or a guarantor against counterparty risk. In fact, counterparty risk doesn't exist; payment is made as soon as the block has been confirmed. In addition, transactions are irreversible (once a sufficient number of confirmations has been reached), because the blockchain prevents changes to the ledger. This is of course fundamentally because blockchain has been devised to simulate cash payments.

Does this signal an end to conventional banking and an irreversible rise in the everyday use of cryptocurrencies? That looks unlikely in the medium term, since it would have great consequences towards liquidity and cost of setting up the system; banks will definitely not adopt ledger-sharing systems, which would mean they could learn a lot about their competitors' businesses. Also, it's difficult to imagine banks willingly delegating control of their record-keeping mechanisms to miners or any other third party. Banks currently are deploying blockchain to suit their needs while maintain a tight control over their operations. An example of the latter is J.P. Morgan,[11] for example, which revealed in early 2019 plans to pilot JPM Coin [14], its own US$-linked stable coin. Unlike other stable coins, such as Tether, JPM will only be used internally for instantaneous transfers between accounts.

Looking to the future, J.P. Morgan is, together with 250 other banks, piloting a blockchain-based Interbank Information Network (IIN) [14]. Other banks, notably UBS, Royal Bank of Canada, the National Bank of Abu Dhabi and Santander are using existing networks, such as Ripple, which enables them to confirm and legally validate cross-border payment details in real time, prior to initiating a transaction through a 'permissioned ledger'.

Cryptocurrencies are much more likely to strengthen the role they play in facilitating low-value transactions in cross-border payments, as an alternative payment network, in countries where banking infrastructure is unreliable or non-existent, and where there is a strong demand for micro-financing. We can see how cryptocurrency can be used to transfer funds by returning to our friends Bob and Alice. Globetrotting Bob wants to send money from Angola to Alice who is in Turkey using his cryptocurrency of choice, bitcoin. He exchanges local currency into bitcoin, which he sends to Alice's public address. Alice exchanges the bitcoin into Turkish Lira. Companies like BitPesa[12] are successfully using this method to deliver cross-border payments or remittances within hours with average fee between 1 and 3%. So long as there are cryptocurrency exchanges at both ends of the transaction, the system removes the need for banks, unlike conventional money transfer systems.

---

[11] See https://www.bbc.co.uk/news/business-47240760 (accessed April 26, 2019).
[12] See https://www.bitpesa.co/ (accessed July 14, 2019).

This system has been so successful that established banks are starting to use blockchain for cross-border payments, and not just in the developing world.

From this, it's possible to see the significant inroads blockchain is making in the conventional way in which payments are made and financial institutions operate. The two most obvious and important examples are:

1. By providing alternative payment infrastructure, where standard banking infrastructure is poor or doesn't work
2. By integrating with current bank infrastructure, as with Ripple, to deliver a more efficient and effective service.

The rapid expansion of mobile technology is, arguably, a prime reason why cryptocurrencies have become a popular payment method, and in all probability will continue to be so. In isolated rural communities, the local bank business model has never worked well, and Telecom companies have filled the gap, deploying mobile technology to transfer money between accounts.

M-Pesa [16] (M standing for mobile and Pesa being the Swahili word for money) was launched in Kenya in 2007 is an example of a successful mobile operator that offers financial services. Today, M-Pesa users can deposit, withdraw, transfer and loan money easily and cheaply via their mobile device; a local agent enables the transfer from cash into digital money. M-Pesa, a banking service with no brick-and-mortar branches, is successfully regulated by Kenyan Financial Service, providing an example for other regulatory authorities considering how to keep up in a rapidly changing financial world.

M-Pesa does not deploy blockchain technology; its innovation relies on the deployment of low-value payments on the mobile phone for the developing countries (Tanzania, Afghanistan, South Africa, India, Romania and Albania) where there is a large population of unbanked citizens in need of financial services

New technologies, like mobile payment, have penetrated the market in the developing world and have the potential to lift people out of poverty. Whether cryptocurrencies will play a large part in helping a new generation of financial services remains to be seen.

Summary

As we have seen, the number of cryptocurrencies has exploded since bitcoin made its debut in 2009. They can be an investment or used to pay for goods and services. The majority of users are individuals, but hedge funds and specialised crypto-funds have started to enter the market.

However, you view cryptocurrencies, when they are used to pay for things, users need carefully consider the benefits of operating outside the banking infrastructure against the cost of exchange and miners' fees.

For low-value cross-border payments, cryptocurrencies can make a difference. What is less clear, however, is whether the benefits can be replicated in the domestic market. In the developing world, domestic payment systems are fast and hardly bear fees. In the developing world, mobile money is also delivering fast payments for low fees.

Interestingly, conventional banks are also piloting with blockchain technology to improve the time and the cost of cross-border payments for businesses.

Frequently Asked Questions

Can I use any cryptocurrencies to pay for goods and services?
In theory, yes. In practice, it depends on how widely accepted cryptocurrencies are. Popular cryptocurrencies that are accepted in many countries are: Bitcoin, Ether, Litecoin, Dogecoin (mostly for micro-tipping people who create content) Bitcoin Cash, Dash, Monero, Ripple and Zcash.

What's the long-term future for crypto-assets? Take-up of cryptocurrencies depends on:

- volatility
- market capitalisation
- number of daily transactions.

Today, more than ten years after its introduction, Bitcoin has market dominance in an increasingly crowded field. Why? Because people like using it!

Why can't I still trust my bank?
No one is saying banks can't be trusted. However, recent revelations of privacy breaches by Silicon Valley giants, such as Facebook, have made privacy (and violation thereof) a hot topic. Most people care about their privacy, some a lot more than others.

If private coin ledgers can't be easily read, how can we know people are cheating?
Cryptography enables the sender and receiver to identify themselves to each other. Apart from them, no one else can view transactions. Miners can validate transactions without knowing the details.

Are private cryptocurrencies more likely to be used by cybercriminals?
Most private coins are used for legitimate business. However, research has shown that in some cases private cryptocurrencies have been used in illegal transactions.

What are the risks of buying cryptoassets?
The greatest risk is that any of these currencies can be worthless in very short period of time. If you have any doubts, don't spend more on cryptocurrency than you can afford to lose.

In terms of investment, how long should I keep a cryptoasset?
While we are not qualified to provide financial advice, it's sensible to consider the spread in exchanging cryptocurrencies can be very high compared to fiat currencies. So, if you are looking to make a short-term profit, beware of the somewhat wild price fluctuations.

Is there a cryptocurrency credit card?
Yes, there is a full financial ecosystem that is being developed for cryptocurrencies, including credit cards and crypto-loans.

6

When and Why to Use Blockchain

Hopefully, you will now have a better understanding of blockchain technology and the advantages it has over traditional payment systems. It's now time to look beyond the relationship between blockchain and cryptocurrencies and examine in detail ways in which this innovative technology can be used in other areas of commerce. Bitcoin is a very good and widely understood example of a blockchain, where anyone can participate: it is what's known as a *permissionless blockchain*. Over the last few years, an alternative has emerged, where participation is tightly controlled; it's called *permissioned blockchain*. We will be looking at both later in the chapter, when we will see how they support a trustworthy, power-sharing way of conducting business. Understanding permissionless and permissioned blockchains is essential to help businesses choose which one is best for them. Later, we will look at some case studies which highlight the benefits and risks of blockchain technology away from its well-publicised role as a platform for cryptocurrencies.

Who Is Using Blockchain?

According to research,[1] by 2022 global spending on blockchain solutions could reach $12.4 billion, a figure that doesn't include cryptocurrency capitalisation or investments in new ventures provided by Initial Coin Offerings

---

[1] See the Worldwide Semi-annual Blockchain Spending Guide from International Data Corporation (IDC).

© The Author(s) 2020
M. G. Vigliotti and H. Jones, *The Executive Guide to Blockchain*,
https://doi.org/10.1007/978-3-030-21107-3_6

(see Chapter 9). What is abundantly clear is that businesses around the world are investing heavily in blockchain as a way of improving their services.

Given the use of cryptocurrencies as an investment and blockchain in cross-border payments (see Chapter 5), it's not surprising that the financial sector is leading the way in blockchain investment.[2] Within the sector, blockchain is also used for regulatory compliance, custody and asset tracking and trade finance.

What, then, can blockchain provide businesses that other more established technologies can't?

Blockchain as a Problem Solver

The distributed ledger might sound like something from a Dickens novel, but it's actually a very modern technology which enables trustworthy timestamping of documents.

World Bank Bonds

An example where blockchain helps with settlement of bonds is the World Bank, which is using blockchain to sell a bond which will be known as Bondi [42], or *Blockchain Offered New Debt Instrument.* The World Bank raised $100 million in issuing the Bondi.

For an example of how it works, let's return to our old friends Alice and Bob who are now about to become business partners. They sign a partnership contract, scan it and both keep digital copies. Unfortunately, this is not the best way to go about things. Why? Well, while they trust each other, either one could, in theory, breach that trust by secretly drawing up a new more favourable contract, and then:

1. Add the same date.
2. Fake the other person' signature.
3. Scan the fake contract.

Were that to happen, the aggrieved party would have to prove possession of the original copy of the contract, wasting time and costing money. Had

---

[2]According to PwC, 82% of use cases for blockchain were in the financial industry in 2017, but 2018 has seen a broadening out of use cases, with only 46% related to financial services—see https://www.pwc.com/gx/en/issues/blockchain/blockchain-in-business.html (last accessed 17 March 2019).

they used a blockchain, they would have digitally signed the contract and inserted it on the ledger.[3] This would allow them to show unequivocally that the copy of the contract, held by each of them, existed when it was registered. The opportunity to falsify and dispute the contract is greatly reduced. Bob could create an alternative copy of the contract, but he would not be able to prove that the new version of the contract is the same as the one recorded on the blockchain. Having realised that this problem is rather common, Alice, Bob contact their professional circle, Ted, Mary, Sunil and Olga and set up a blockchain so that they can safely sign multiparty contracts.

Boldly Creating Trust Where None Has Gone Before

Creating trust among competing companies—never an easy task—is one area where blockchain really comes into its own.

To show how this can be done, let's take a trip to the movies. Popcorn at the ready, sit back and enjoy 'Star Trek VI: The Undiscovered Country', where interstellar suspicion and skulduggery threatens disaster. Still boldly going where no one has gone before, Captain James T. Kirk and Dr. Leonard McCoy of the USS Enterprise are wrongly accused of launching an unprovoked torpedo attack on a Klingon vessel. Tried and found guilty by the Klingons, both men are sentenced to life imprisonment on the penal colony Rura Penthe, a frozen planetoid, where life expectancy for a prisoner was one year.

Despite their protestations of innocence, things look bad for Kirk and the long-serving doctor when the ship's log appears to confirm that the torpedoes had indeed been fired by the Enterprise. It's left to Mr. Spock, the Enterprise's science officer and master of logic, to prevent an icy outcome for Kirk and McCoy by showing that the log had, in fact, been altered with collusion from another alien-race well known to Star Trek fans, the Vulcans.

At this point, readers may have two questions in mind: firstly, if it's that easy to change the log on the Enterprise, isn't it time for Kirk to order a thorough cybersecurity review? And secondly, why is this movie detour, entertaining as it may be, included in a book about blockchain?

---

[3]We assume that Bob and Alice do not mind other people reading the contract.

The first question will have to remain unanswered, but the second is, fortunately, straightforward. If blockchain had been thought of by the scriptwriters, the movie would have been much shorter, considerably less entertaining and a box-office flop. Had a blockchain been in place, all the protagonists in the movie would have maintained a ledger (see Chapter 4), which would have included a log for the Enterprise and other vessels. Log owners[4] would insert a cryptographic digest on the blockchain, which need to be both timestamped and digitally signed by 100 independent ships, randomly chosen from the Blockchain Starfleet consortium. The ledger of the Blockchain Starfleet is a cryptographically signed chain of records belonging to vessels whizzing across the universe. If the Enterprise had wanted to fool the blockchain and change a transaction, it would have had to ask all 100 ships forming the blockchain consortium to agree to resign the transactions, and change all transactions inserted afterwards. That's about as likely as the heroic Captain Kirk resorting to such underhand tactics in the first place![5]

So, we can see from this that a blockchain can create trust, even among parties whose first instinct may be to mistrust each other. It can also help avoid costly litigation and provide transparency and security.

When Blockchain Isn't the Answer

Believe it or not, blockchain isn't the solution to every problem. It can't, for instance, prevent an insider threat. If a rogue engineer on the Enterprise had manipulated the ships log before its insertion in the Starfleet blockchain in a dastardly attempt to implicate the Enterprise, the blockchain would have become instrumental in the success of the Klingon's evil plan.

From this, it's possible to see that a blockchain behaves in much the same way as any other system. There's a memorable phrase to describe this: Garbage In, Garbage Out (GIGO). The quality of data on the blockchain is crucial for it to function efficiently.

---

[4] Most of computer systems have logs or log files, where all important events are recorded that occur during the running of a computing device; it can be used to understand the activity of the system and to diagnose problems.

[5] Of course, it would have been equally impossible for him to cheat on the Kobayashi Maru scenario (by secretly reprogramming the simulation computer), which would have likely stymied his career in Starfleet.

Centralisation Versus Decentralisation

Decentralisation is an important feature of some, but not all blockchains, including cryptocurrencies, and it can mean: *share of power* or *disintermediation.* There are two ways in which decentralisation comes about:

Technical decentralisation Allows a site's technical backbone—known as a node—to stop working without fundamentally affecting the system as a whole. All nodes have very similar functionality, which prevents one node from becoming more important than another (share of power). Examples of this kind of infrastructure are BitTorrent, the original Spotify,[6] and the earliest version of Skype.[7]

Governance decentralisation No company or institution is in charge of operations (disintermediation). Taking Skype as an example, the company was centralised, but deployed a decentralised network to efficiently run its operation.

The most obvious example of technical decentralisation is the Word Wide Web (the Web), the global information medium invented almost thirty years ago by Tim Berners-Lee. Today we have a space where, thanks to the Internet of course, anyone can set up a webpage. There is no overarching authority controlling or steering the web, other than a consortium, W3C, which manages interoperability standards.

In the case of Bitcoin, Ethereum or other cryptocurrencies, decentralisation comes in the two forms mentioned above: technical and governance. Technically, if all Bitcoin nodes stopped working, then Bitcoin would cease to exist. However, because the Bitcoin community has significantly expanded since its inception, if a few nodes were lost, they would quickly be replaced by others wanting to join the network.

---

[6]Since 2014, Spotify does not use peer-to-peer networks anymore.

[7]In the last ten years, Skype has moved away from peer-to-peer network.

Peer-to-Peer Networks

Peer-to-peer networks denote a group of connected devices that are more or less equal in functionality, as oppose to traditional networks where there is at least one vital coordinating device without which the network cannot fully operate.

BitTorrent is an example, notorious for being used in illegal downloading, but that association doesn't do justice to the benefits of these networks; BitTorrent's main purpose is to make transferring large files, like the source code for the Linux operating system, between many people easier.

As for managing truly decentralised cryptocurrencies, there isn't one single authority, company or enterprise directing or controlling operations. Indeed, there are no objectives nor targets, nobody can fire miners or people running nodes and miners can join the network without permission. Software upgrades, which follow clear rules agreed by the cryptocurrency's community, are carried out voluntarily.

Developers and miners occasionally meet (at conferences, for instance) but communications normally take place online.

What is it, then, that makes decentralisation so appealing? Firstly, *distributing power among participants* protects all their interests. When a project like Bitcoin, Ethereum or cryptocurrencies is fully decentralised, ideally[8] not one voice has more of a say or greater influence than others. Crucially, it also makes it impossible for Bitcoin to be closed down by governments or large organisations.[9] For that reason, fully decentralised cryptocurrencies or blockchain are often called *uncensorable*. Of course, that doesn't stop governments from effectively blocking a cryptocurrency by simply making payments or running a node illegal.

The original goal of decentralisation, for bitcoin at least, was to enable cheaper electronic micro-payments [94]. It's unimaginable that anyone would use bitcoin if it was regarded as vulnerable to fraud; decentralisation is seen as a reliable way of preventing that from happening.

A fully decentralised, permissionless public blockchain is not the solution for all, and would have alarm bells ringing loudly at most companies, since they would prefer to keep the governance centralised. But technical decentralisation

---

[8]There is a debate as whether Bitcoin is fully decentralised with so few miners left/at the moment.

[9]It would be possible for a state nation to attack or shut down some part of the Internet connections and isolate, for example, a country or a continent. For simplicity, we exclude these kinds of scenarios that would have repercussion on functioning status of the Internet as well.

Technical Centralised Ledgers	A Distributed Blockchain Network
Any participant trusts that somebody is properly backing up the data.	Creates as many backup copies as the number of participants; all copies are continuously updating and syncing to the same ledger data. Each participant can maintain his/her own copy of the ledger making loss or destruction of the ledger difficult. [a]
Robust resiliency could be difficult to achieve: if the infrastructure replicates similar components an attack on one part of the network will work everywhere.	Participants are free to deploy compatible software, hence nodes on the network can be different. An attack on one node would not work on other nodes, if they deploy different components.
Any participant must trust that the owner is validating each received transaction. The transactions are not made transparently and may not be valid.	All transactions permanently recorded on the ledger are valid; invalid transactions are prevented from propagating throughout the blockchain network. [b]
The transaction list on a centrally owned ledger may not be complete; a participant must trust that the owner of the ledger is including all data, and that transactions have not been modified. [c]	A blockchain ledger keeps all valid transactions. Any new block references the previous block. [d]

Fig. 6.1 Comparison centralisation versus decentralisation ([a]Some blockchain implementations provide the capability to support concepts such as private transactions or private channels. Private transactions facilitate the delivery of information only to those nodes participating in a transaction and not the entire network. [b]The assumption is that the probability of a 51% attack is very low. [c]See Chapter 4, section 'What Is Blockchain Technology?'. [d]See Chapter 4, section 'How Does the PoW Work')

delivers some great benefits as outlined in Fig. 6.1 [108]. Major tech companies like Facebook, Google, Amazon and LinkedIn operate under a governance centralisation, regardless of how the technical infrastructure is organised. The benefit is that the company has full control over every single part of operations and any associated risk can be managed and monitored. Control, however, is achieved at a cost. The old Skype, before acquisition by Microsoft, was able to deliver good services at low cost, with millions of users. However, some serious technical problems came to light, and in 2009, Skype decided to leave out the peer-to-peer technical solution because it did not fit with the use of the app on mobile phones.

There is a growing interest from enterprise in blockchain as the technology enables an approach in using both distributed ownership as well as a distributed physical architecture. Is there, then, a way of achieving technical and commercial decentralisation that delivers the benefits we have outlined here while at the same time reducing the perceived risks and lack of control associated with permissionless blockchain? That's what we will be looking at next.

Blockchain for Enterprises

Some enterprises need to share data with lots of multiple parties and may need to update common information while each of them wants or needs to maintain control, and supervise what changes have been made by others. This happens in sectors where compliance is crucial, and parties have to fulfil a specific role. A common concern is that a mistake by one party can have an adverse impact on another. In such cases, each participant maintains a local ledger which, inevitably, differs slightly from the rest of the group and considerable effort is needed to ensure that everyone has the same view and that end-to-end transparency is maintained. In cases where companies have separate boundaries, ensuring that there is still a common view requires time and effort; there are situations where, to prevent disputes or cheating, an independent third party might be required to foster an environment free of doubt or suspicion. Blockchain technology's goal is to produce a set of authoritative records that are validated and executed by multiple separate entities via a consensus process that dispenses of central authority [60].

A blockchain displays *Key Blockchain Properties* (KPB); namely, the ledger is a *distributed store of records* organised in *chronological order*, such that it:

Isn't owned by a single organisation For example, Bitcoin (or other cryptocurrencies) transactions resemble a bank statement. Unlike a bank, however, there is no single organisation authorising or blocking transactions. No organisation owns Bitcoin, and no single organisation can shut it down.

Requires a consensus process Usually, consensus leads to better decisions. In business, decisions reached by consensus are recorded by human beings, whereas a blockchain consensus is reached and recorded by an automatic procedure (an algorithm).

Can be updated with append-only operations A blockchain ledger can only grow by inserting new transactions. In the Bitcoin blockchain, all confirmed transactions can be seen and retrieved, but never modified.

It's tamper-proof and cannot be overwritten The blockchain consensus has been devised to prevent sophisticated attacks that could modify past (confirmed) transactions.

> ## Benefits of Blockchain to Enterprise
>
> It's not overstating the case for blockchain to say that it has potential to fundamentally change the way some businesses operate.
> Blockchain, similar to other technologies, provides the following advantages for modern business:
>
> **Transparency** A distributed ledger is the only source of truth for blockchain participants.
> **Scalability** Great number of transactions recorded on the ledger can drive businesses.
> **Digitalisation** There's no need for lengthy paper-based processes.
> **Completeness** All data is stored on the ledger.
> **Elimination of third parties** Participants can interact directly and safely with the platform—no need of third party to validate the data.
> **Tamper-proof data history** Data on blockchain can be verified by every participant, to the highest standard of compliance, as it is tamper-proof.

There are several reasons why most companies would not deploy permissionless blockchain.

1. A public permissionless blockchain has its own well-defined goals, which rarely align with those of a company.
2. They are run by an amorphous community with no legal status in many jurisdictions. For example, companies that need to release software upgrades in response to new regulation would find it challenging to rely on a community to deliver such a critical task.
3. Some suffered from scalability issues: Bitcoin and Ethereum settle only a few transactions per second, while businesses could need a few thousand.

A permissioned blockchain is created when participants inserting data into the ledger keep a tight control over who can participate, while the ability to read data remains public. A private blockchain is one where authorisation is required to read the ledger, and is suitable in case confidentiality of the data is the primary concern, for example in military applications. Permissioned blockchains deploy a distributed ledger and deliver trustworthy power sharing for a limited number of carefully selected participants that form a *consortium*. The consortium, made up of participants who have a common goal, often operates under the umbrella of a common agreement. To operate efficiently, this type of blockchain requires the consortium to agree on a set of well-defined *governance rules*.

As a first step, there is the need to identify the participants of the consortium and determining:

1. The goals and objectives
2. What data needs to be shared: only data needed to address a problem that is common to the consortium needs to be shared (no distiction is made between the 'hash' and the 'metadata')
3. The format of the shared data
4. Which parties will be *validators* of data (permissioned blockchains don't normally need miners)
5. What kind of consensus process is going to provide a sufficient degree of security for the ledger.

In this kind of blockchain, there is the freedom to decide what kind of consensus is used. The PoW assumes an open adversarial environment: anybody could be potentially cheating. When a blockchain consortium is established together with the validators, the consensus process is likely to be less stringent, as the risk of cheating is reduced.

Any consensus process should have the following properties to derive the maximum benefit from a blockchain:

1. All participants/validators must agree on what information is added to the ledger.
2. Final information appended to the ledger should be produced by an honest participant—anybody trying to cheat should not be able to succeed.[10]
3. The ledger must be shown to be tamper-proof under the chosen consensus process.

Despite the checks and balances inherent in a permissioned blockchain, fraudulent activity, associated with participants colluding, can't be entirely ruled out. The bigger and stronger the consortium, the easier it is to devise a

---

[10]This may appear trivial to say, but it is very difficult to define a procedure that is actually secure.

robust consensus process to ensure that everyone abides by the rules. Private or permissioned blockchains with a very limited number of participants have their critics, not least because it would be very difficult to claim the consensus process is instrumental in preventing parties cheating. Even worse, as the blockchain is designed to operate with multiple parties, deploying a private blockchain for one single company would not be a good blockchain application. There are better cryptographic tools, that are not blockchain, that can help with a small group of participants in need of protecting their data.

Permissioned blockchains can be scalable depending on the consensus process chosen. This is one of strengths of the permissioned model. It's crucial that any consensus is carefully evaluated before setting up any blockchain, and a trade-off must be found between scalability and security. If participants do not need a consensus process, then the question is really: Do they need a blockchain? In the next section, you will find a useful description of a platform enabling a group of participants to work together and share data, which is not a blockchain.

The benefits and risks of the types of blockchain are shown in Fig. 6.2. If you would like to know if a digital system could be a blockchain, a simple way to find out is to follow the questionnaire in Fig. 6.3.

Round-Robin Consensus Model

Several examples of consensus mechanisms for permissioned blockchain can be found in the literature [108]. The Round Robin Consensus is a way to ensure that nobody inserts more transactions than others on the ledger. It would be useful in a case where there is a need to guarantee equal participation among the member of a consortium. Participants take turns, following the 'Round Robin' strategy, in creating digitally signed blocks on the ledger, which are linked together to form a chain (see Fig. 4.9, section 'How Does the PoW Work?'). This operation needs to be completed within a pre-agreed time window, so that if a participant is not available to insert the transactions, the system will not halt. To ensure that the ledger is tamper-proof, crucial information about the links among the blocks is published on national newspapers around the world on a regular basis [83]. Only participants whitelisted can take part in the consensus process; otherwise, fake participants can be created to subvert the original goal.

		Public Ledger	Private Ledger
Permissioned		Selected participants form a consortium to determine: • Rules to validate the blocks on the ledger • Participants validating the transactions • Participants allowed to submit the transactions • Governance agreements	Selected participants form a consortium to determine: • Rules to validate the blocks on the ledger • Participants validating the transactions • Participants submitting the transactions • Governance agreements
	Ledger	Can be queried by anybody	Can be queried only by a pre-approved set of participants
	Benefits	Control over partners joining the consortium Efficient validation of blocks with opportunity to create scalable platforms Reduced legal and accountability risks	Control over partners joining the consortium Efficient validation of blocks with opportunity to create scalable platforms Reduced legal and accountability risks Handling confidential and private information
	Risks	Costly to set up and customise Potentially unsuitable for confidential / sensitive / private information	Costly to set up and customise
Permissionless		No need for a consortium of participants. Anybody can validate block of transactions	
	Ledger	The ledger can be queried by anybody	
	Benefits	Blockchain infrastructure available Easy to run pilots and applications Full transparency of the transactions	
	Risk	Unsuitable to deal with confidential / sensitive / private information No control over the future development of the blockchain Lack of legal and accountability framework	

Fig. 6.2   Risks and benefits of permissionless and permission blockchains

It Looks Like a Blockchain but …

So far, we have looked at what a blockchain is and how it can deliver transparency, cost savings and efficiency for business. Now we need to spend a short time looking at the difference between new smart technology and blockchain.

Submitting VAT and end-of-year accounts is a familiar and necessary task for anyone running a business. It's not exciting, but it has to be done and for many years the process remained almost unchanged. A company would:

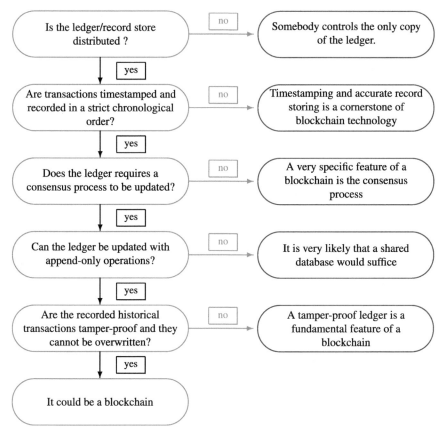

Have you answered 'no' to at least one of the questions above?
Then you know it is not a Blockchain

Fig. 6.3 Find out if it is not a blockchain

- Record expenses on a spreadsheet with scanned or digital receipts.
- Record invoices issued with the associated scanned or digital official document.
- Calculate costs, VAT and potential returns.
- Pass records to the company accountant, with contracts, bank statements and other relevant documents.

It would then be the accountant's duty to check documents for accuracy and provide final figures in compliance with the law. They would, of course, keep a copy of the submission in case of further investigations by revenue and customs department.

Thankfully, in recent years new cloud-based software has emerged to stream-line what was a time-consuming and repetitive process. Now, thanks to this clever solution, a business and its accountant can easily and painlessly share financial transactions needed required by the tax authorities.

This enables a business to:

- Automatically connect with the bank and download statements.
- Issue, record and backup invoices.
- File expenses, link them to both transactions shown on bank statements and to scanned or digital receipts.
- Correctly label expenses in compliance with tax rules.
- Record and backup company-wide expenses.
- Compute the total of expenses, profit, VAT and any other taxes due.
- Ensure information is in chronological order.

The accountant can see all this information, saving time and making the process much more efficient. To comply with the law, he or she simply needs to:

- Verify the labelling.
- Ensure expenses are backed up with receipts.
- Ensure compliance with the law.
- Click to compute VAT and TAX.
- Provide confirmation that his/her understanding is correct.

Collective Random Signatures

This is the consensus described earlier in the section 'The Star Trek: Boldly Creating Trust Where None Has Gone Before'. It would be useful to provide extra security that past blocks cannot be unilaterally modified by a single participant, but to work, it assumes a rather large consortium. In this model, blocks can be inserted by any participant: provided that they are timestamped and digitally signed by a third of the consortium, which is randomly chosen for every block. The ledger will form a linked chain of records (see Fig. 4.9, section 'How Does the PoW Work?') To override a past transaction need the collusion of all co-signatory participants for each block attached after the transaction in question: in brief links among the blocks ledger needs to be recreated with the help of the co-signatory participants.

A combination of the cloud and some very helpful software has made accounting a speedier process these days for a lot of firms. The revenue and customs department is also happy because it's easier for them to inspect a company's accounts and see the link between digital records of expenses and invoices and outgoings and money coming in. It's a win-win.

While accounting systems (AS) described above enable multiple parties to share, and access and update sensitive information that has to be correct, and offer an audit trial useful for compliance purposes, it looks like a blockchain, but it's not, for the reasons shown in the table below.

General Property	AS	Explanation
Is the ledger/record store distributed?	No	One ledger connects the company members, accountant and cloud provider and, in the future, the tax-man
Are transactions timestamped and recorded in a strict chronological order?	Yes	Transactions are: invoices, expenses, reports, and bank transactions imported from the bank
Does the ledger require a consensus process to be updated?	No	Merely requires company accountant to insert transactions and the bank to download the data
Can the ledger be updated with append-only operations?	No	Wrong invoices can be deleted
Are the recorded historical transactions tamper-proof and they cannot be overwritten?	No	Old invoices / expenses can be unilaterally changed, if, for example, there was mistake

From this example, it's possible to see that there are other technologies which, like blockchain, can save time and money, provide transparency and deliver efficiency, but that do not deliver the unique benefit that belongs to blockchain: tamper-proof data equally managed by a group of participants. This feature makes the blockchain a trustworthy power-sharing technology.

Case Studies

Below are a few wide-ranging examples of where blockchain is successfully deployed by companies around the world.

Timestamping Documents

As we saw with Bob and Alice's contractual partnership, blockchain can be used for timestamping and recording valuable documents, in a way that enables people to show that documents were not changed at a later stage and exist from a specific point in time. Documents can be recorded and timestamped for a number of reasons, including:

Value Certificate of land, academic certification, property, gems etc.

Contracts Contracts, share allocations, etc.

Records of transactions For example, to continue to pay a low premium on a life insurance, you may need to show you regularly go to the gym; each visit can be recorded on the blockchain together with details of the physical activity undertaken.

One common misconception about blockchain is that it can only be used by multiple parties. This is not the case: in fact, a company can use an existing blockchain simply to timestamp documents or relevant information. Open permissionless blockchains such as Bitcoin or Ethereum, and many others, would support such functionality without the need to form a consortium. This would be suitable in the case where a single organisation or individual needs to act alone, for example for proving that ideas, stories or agreements existed from a specific point in time.

How would this solution work without divulging too much information? The key point is to appreciate that only a unique identifier, known as a cryptographic digest (see Chapter 4), needs to be stored on the blockchain, not the document itself.

The user can insert this identifier on a blockchain as any other transaction, pay the miners' fee and keep the original document secret. From the unique identifier, it is very difficult to construct the original document, and hence, there is no need to worry about other people stealing the information. When the time comes, to prove to third parties that the document existed at a specific time, only what is needed is to retrieve the cryptographic digest from the blockchain (one needs to remember in which block it was stored), recreate the cryptographic digest for the document and show that the digest on the blockchain and the digest from the document are the same. Because the blockchain is tamper-proof, and cryptographic digest is unique to the document, this is proof the document existed from the time it was timestamped and registered on the blockchain.

If there is a need to operate on a consortium, then there is a decision to be made which blockchain is best. In the presence of several parties, with different needs, permissioned blockchain can be easier to deploy, but there is an upfront cost in setting up the infrastructure, providing access to the right partners, to organise the governance and to agree on what kind of consensus process should be adopted. Deploying a permissioned blockchain for merely timestamping data could be overkill.

Supply Chain Management

Finding the correct fit for your organisation is important when choosing a blockchain.

A permissioned blockchain could work well for a business operating in global logistics and supply chain management. The life cycle of any type of product, from production to consumption, requires documentation that can be trusted. Each link in the supply chain is required to make a careful record of the product. It almost goes without saying that a reliable and tamper-proof history of a product is vital to everyone involved in the chain. This is where blockchain comes into its own, keeping accurate records, tracking assets and ensuring that everyone in the supply chain maintains control of the data.

The typical list of stakeholders can be found in Fig. 6.4, and their role in the blockchain is listed in Fig. 6.5.

We see a decentralised permissioned blockchain where each participant, apart from the end-user, runs its own node (Fig. 6.6). Governance is through

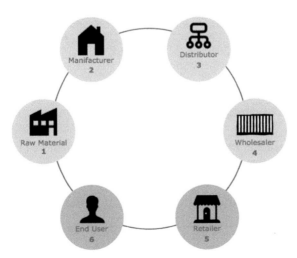

Fig. 6.4 Supply chain stakeholders

Deployment of Blockchain in the Supply Chain	
Stakeholders	Producers of raw materials, manufacturers, distributors, wholesalers, retailers and end-users
Content of transactions	Time stamped, status of the product, phase of the supply chain reached
Validators	Producers of materials, manufacturers, distributors, wholesalers, retailers
Ledger Accessibility	Everyone including end-user
Candidate for Consensus Process	Random co-signature to prevent changes in data
Governance	Contractual obligations to participate and insert correct data, group of participants manage membership of blockchain
How new members join the consortium	Via request to group that manages blockchain platform
How members leave the consortium	Via request to group that manages blockchain platform

Fig. 6.5 Blockchain supply chain table

General Property	Permissioned Blockchain	Explanation
Is the ledger/record store distributed ?	Yes	Each stakeholder runs its own node, and stakeholders share information about the product to be delivered
Are transactions timestamped and recorded in a strict chronological order?	Yes	See Figure 6.7
Does the ledger requires a consensus process to be updated?	Yes	Co-signature of each transaction inserted on the blockchain to prevent tampering
Can the ledger be updated with append-only operations	Yes	All the transactions are recorded
Are the recorded historical transactions tamper-proof and they cannot be overwritten?	Yes	Not as secure as some public permissionless blockchains

Fig. 6.6 Key blockchain properties in the supply chain

a representative subgroup of stakeholders, who also manage the consortium's membership.

An overview of the stakeholder transactions is shown in Fig. 6.7. The consensus process can be decided: we suggest as a consensus the random selection of co-signatures to prevent easy overriding of the historical transaction of the ledger and a seizable number of transactions per second to be inserted on the blockchain.

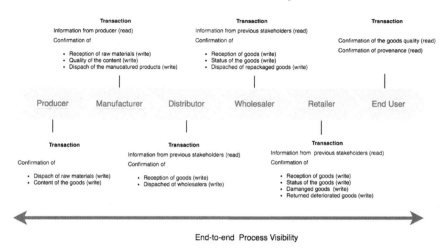

Fig. 6.7 End-to-end process visibility

The above example could be adapted to cover complex, global supply chains for valuable goods such as precious gems or, even better, to facilitate the prevention of counterfeit drugs.

A final word of caution: as discussed in section 'When Blockchain Isn't the Answer', the blockchain itself doesn't prevent bad data to be added. It does have the potential to make it harder. The end-to-end transparency provided by recording the journey of a product from manufacturer all the way to the shelves of the shops makes it harder for somebody to record wrong or false information. In the case precious gems, if false information is recorded and later discovered, it would be relatively easy to determine and investigate the party who has introduced the wrong information.

Trade Finance

As the name aptly says, trade finance is the finance of trade that enables goods and services to move around the world. In 2018, the global trade finance market was at around $6 billion [10], and it is expected to grow to $7 billion by 2024. What are the challenges in trade finance? Starting from the beginning, a trade transaction requires a buyer and a seller. In international trade, the shipping of goods can take months, exposing the parties to various risks: the seller risks of not being paid after delivery of the goods, and the buyer risks of receiving wrong or damaged goods. If things go wrong, both parties face the added challenge of having to recover the payment in a foreign and unfamiliar legal and political system. For small and medium companies, trade finance can be a minefield, and it is often an obstacle to trade.

There are four common payment methods in trade finance:

Cash in advance The buyer orders and pays for goods before shipment is made, and the seller ships goods and associated documents to release the goods (Invoice and Bill of Lading) upon receipt of funds.

Open account The seller ships goods and sends associated documents, and the buyer pays for goods upon their receipt

Documentary Collection The seller ships the good, but the documents are held by a third party, typically a bank; the latter sends them on to buyer's bank for the release of good, provided the payment terms are met.

Letter of Credit A letter of credit is an official document released by a bank, which assures the seller that payment will be made provided the agreed terms specified in the letter of credit are met.

A letter of credit provides both the buyer and the seller a peace of mind (see Fig. 6.8). Letter of credits are only used by a small fraction of participants as they can be costly and time-consuming and often, the end-to-end process of deploying a letter of credit is not standardised. This type of trade finance involves the exchange of many documents between the buyer, the seller, their respective banks and agents making physical checks of shipped goods at each checkpoint, as well as customs agencies and freight insurers. This leads to delay in production, which in turn can have an impact on the liquidity of funds or cash flow, which is a big problem for small and medium enterprises.

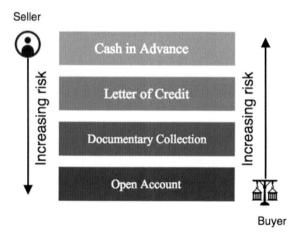

Fig. 6.8 Common payment methods in trade finance

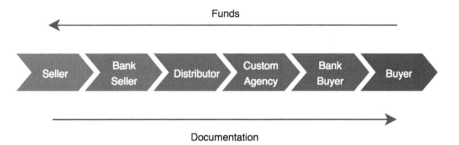

Fig. 6.9 Stakeholders in trade finance and description of the blockchain solution

Deployment of Blockchain TradeFinance	
Stakeholders	Buyers, Sellers, Banks, Financial institutions, Distributors, National Custom agencies
Content of transactions	Time stamped documents such as: letter of credits, bills or lading, status of the shipments, compliance national law for import
Validators	Banks and financial institutions, Distributors, National Custom agencies
Ledger Accessibility	Only the stakeholders
Candidate for Consensus Process	Random co-signature to prevent changes in data
Governance	Contractual obligations to participate and insert correct data, group of participants manage membership of blockchain
How new members join the consortium	Via request to group that manages blockchain platform
How members leave the consortium	Via request to group that manages blockchain platform

Fig. 6.10 Blockchain for trade finance

Letter of credits exists for hundreds of years, and they are well understood in the banking system. Yet, the entire process is time-consuming and complex with each party having to make sure agreements are in place. Blockchain can potentially simplify and streamline trade finance process by providing each stakeholder with a reliable history of the transaction (see Fig. 6.9). What kind of blockchain is needed? Similarly to the supply chain example—see section 'Supply Chain Management'—a permissioned blockchain would be preferable to corporates. The ledger would provide an authoritative and tamper-proof history of the import-export process which is available to the parties involved (see Fig. 6.10). This solution would reduce the time of payments for shipped goods to a couple of days instead of a few weeks and increase the volume of trade finance worldwide. Several consortia are piloting blockchain for trade finance, and in future, the blockchain can potentially simplify payments having smart contracts automatically release fund to the seller upon the addition of a valid and relevant documentation to the ledger in real time.

The blockchain would provide a global, reliable trade finance registry—that most participants, including banks, around the world could query, would also help, to prevent seller and exporters to illegally obtain credit multiple times for the same shipment which is a currently a big problem [42].

Compliance

In some sectors, compliance is vital: finance and safety-critical industries, such as oil and gas, nuclear and railways, are good examples. In complex projects, compliance requires the collaboration of several parties and the need to keep the documentation for long periods of time. Broadly, regulatory compliance involves the following processes and blockchain can assist with each one:

1. Recording and storing relevant information, data and documents
2. Aggregating data: required when, for practical or legal reasons, data is stored in multiple systems or locations
3. Performing operations on data: data used in regulatory reporting must often be processed or analysed before being passed to a regulator. For instance, when applying internal financial models to determine compliance with capital adequacy requirements
4. Sharing information with other entities: regulatory reporting obligations require firms to share information securely with regulators, and occasionally with other organisations
5. Ensuring data integrity: firms must have processes to prevent or correct errors introduced by the processes described above.

In these instances, blockchain enables parties to operate and share data by deploying the distributed ledger; because the history of the data is tamper-proof, the blockchain creates auditable history of data that can be easily shared with a regulator.

Land Registry

Land, property ownership and a trustworthy ownership registry are corner-stones of advanced economies. It is estimated that the value of unrecognised land in the world is about $9.3 trillion [103]. Unrecognised land comes about either because there is no paper trail that certifies ownership, or because un-reliable certificates assign the land to more than one owner. Either way, the economic impact is significant: banks, for example, are unable to use un-recognised land as collateral for affordable loans, which in turn could boost businesses. In most developing countries, land registries are managed by a gov-ernment agency, which keeps them up to date and, in most cases, digitalised. In the UK, for example, land titles can be downloaded from a government website for a small fee, enabling anybody to efficiently perform due diligence when buying a property.[11]

These kinds of registries need to be consistent: the same land cannot belong to two individuals at the same time. To prevent this kind of situation, solicitors, operating for the buyer and seller, ensure, among other things, the legitimacy of the title and authenticity of both the buyer and the seller (verifying that they are whom they claim to be). The stipulated government agency will duly execute the transfer by appending, to the registry, a new updated land title. Typically, information on previous owners of a property is kept as part of the registry, to prevent future disputes and for historical purposes.

Reflecting on this, we can see that:

1. Land titles can't be transferred twice from the same owner—unless the land it split.
2. The land registry keeps the history of ownership of each land title.
3. The owner(s) of the land give authorisation for the transfer of the title.

We have seen that cryptocurrencies like bitcoin follow a similar set of rules. This is a prime example of disintermediation of a government agency. In a number of developing countries, government land registry continues to hold paper records, and land certificates are too often forged, sometimes with the collusion of poorly paid civil servants. When deciding who should maintain a land registry, it makes sense to start with banks (see Fig. 6.11) who have a financial interest in preserving the integrity of the registry, and ensuring that the registry operates under strict rules. A land registry on the blockchain can be seen as a sequence of digitally signed land titles. To prevent people cheating the system and transfer land twice, all what the consensus process requires is

---

[11]It is worth noting that also in developing countries land registry blockchain solution is being deployed. For example, in Estonia, the project e-LandRegistry (https://e-estonia.com/solutions/interoperability-services/e-land-register/) is currently operative, and in the UK, land registry has conducted a successful pilot (see https://www.gov.uk/government/news/hm-land-registry-to-explore-the-benefits-of-blockchain). In these cases, the focus is on improving an existing land registry service, as opposed to find a new, fair model to operate.

Stakeholders	Buyer, sellers, banks and other third parties such solicitors or even Government agencies
Content of transactions	Digitally signed land title confirming transfer of ownership
Validators	Banks or Government agencies
Ledger Accessibility	Everyone can query the ledger - it is a benefit to society
Candidate for Consensus Process	Random choice transaction in case of double transfer of titles
Governance	Contractual obligations to participate to check correctness of the titles when inserted for the first time
How new members join the consortium	Via request to the group that manages blockchain platform
How members leave the consortium	Via request to the group that manages blockchain platform

Fig. 6.11 Land registry blockchain table

for the validators to (randomly) choose one of the two new titles and discard the other one.

It's important to keep in mind that land titles are not created in the way cryptocurrencies are—they already exist. Therefore, a permissioned blockchain, with a proper governance, needs to ensure that original titles are correctly inserted, and let the consensus process ensure the transfer of titles. Attributing a title to the wrong person would be clearly wrong, and on the blockchain, it would be difficult, but not impossible, to remedy such a situation. A land registry that works well on the blockchain should also be user-friendly and enable people to retrieve their digital signature in case of loss; enable land to be split between more than one owner; and consolidate tracts of land under one ownership.

Finally, it must be noted that where a land registry does not exist, the blockchain cannot help with the political or social challenge of creating one: all the blockchain can do is implement strict governance rules.

Challenges

There are cases where using blockchain technology, both permissioned and permissionless, can be challenging. Two examples are voting and dealing with personal data.

Voting Systems

At first glance, blockchain might appear to be perfect for storing electronic political voting records: nobody can change the vote, and governments could easily verify the votes preventing long recounts in case of disputes. This is not the whole story. Free voting is the ultimate expression of democracy. We label countries where free voting does not exist as dictatorial for very good reasons and as such, voting should not be rigged. Whether in electronic or standard voting, vote manipulation is prevented by:

Privacy of the vote People should not be obliged to tell anybody for whom they vote, and should not fear any repercussion for their political belief.

Anonymity of the vote Nobody should be able to detect if a particular individual or community has voted for a political party.

A public ledger could easily reveal information about how an individual has voted, even when voting is made anonymous (see, for example, Chapter 4, section 'Querying the Ledger'). This is very important because selling votes is illegal in many countries. If anybody can determine, even if only with a degree of probability, by only looking at the ledger, how a particular individual has voted, then the system would favour selling votes.

There are a lot of sophisticated cryptographic techniques that can be safely deploying to ensure the privacy of the voters [73], and at the same time to enable ballots to be counted. It is not very clear why any voting system would require a consensus: Who would participate in this consensus and why? The person who votes has the sole right to cast the vote, which merely needs to be recorded for the sole purpose to determine the outcome of the election. Once the election is over, the data can and should be ignored. This data does not need to be shared with anybody else beyond those counting the votes.

The only question that remains to be answered is, would it not be interesting to leave evidence for posterity about the voting? The answer is 'No': there is no need to provide any extra information about voting other than the aggregate results of the counting. That is all what a historian needs.

Of course, when it comes to voting for less sensitive causes, for example voting in companies' boards, provided that people agree, these can be stored on any blockchain.

Personal Data and the Blockchain

The European General Data Protection Regulation (GDPR), which came into effect in May 2018 [30], places clear restrictions on how information about EU citizens can be used by companies around the world. Some of these restrictions need careful consideration when a blockchain is used for storing personal information.

When discussing blockchain and GDPR, several issues need to be considered [37]. The 'right to be forgotten', which allows an individual to demand the erasure of personal information under certain conditions, is probably most contentious, because one of the tenets of the blockchain is that the historical transactions of the ledger cannot be overwritten. It might appear that storing clear text personal data that anybody can read and that might need removing in future constitutes a breach of the right to be forgotten.

Of course data on the blockchain can be stored in different forms: for example, encrypted or cryptographic digest could be stored on the ledger of blockchain, without necessarily creating an immediate violation of the right to be forgotten. The GDPR is not prescriptive in the way the personal records are forgotten: therefore, the data does not necessarily need to be physically removed from the blockchain. Other means can be used, so long the record is removed from public view.

GDPR also imposes requirements on data that is transferred outside of the EU. In the case of insertion of clear text data, it would not be possible for an entity to comply with the GDPR if deploying a public blockchain because participants could store a copy of the ledger outside the EU. Again the format of the data stored on the blockchain can be the determinate factor in becoming compliant.

Storage of personal data needs to be carefully considered in all systems, blockchain or not. In what format the data is stored, the sensitivity and value of the data are among the parameters that make a difference between being compliant or not. While it appears that storing data on the blockchain can lead to challenges in compliance with the GDPR, there are ways and means to achieve such a compliance by deploying the right data format.

Summary

As we have seen in this chapter, blockchain is a very useful technology for business, with benefits ranging from a cost-effective method for timestamping documents to co-operating with other parties when it is not possible to share an IT infrastructure. If a public blockchain, such as Bitcoin or Ethereum is not suitable, a permissioned or private blockchain could well be a workable alternative. For these, participation is limited to known organisations that are part of a consortium, ensuring that everyone is working towards a common goal, while enabling each party to keep full control on their data. Key to the success of permissioned blockchain is the definition of the initial governance and the deployment of a suitable consensus process that supports the original business goals. There are obstacles to the long-term development including a lack of standardisation and interoperability. Today, public permissionless blockchains are inevitably more interoperable: there is a more widespread consensus about technology. Permissioned blockchains, by contrast, tend to operate according to the needs of the participants in the consortium. To overcome the first obstacle, in 2016, the International Organization for Standardization (ISO) started an international working group to deliver a blockchain standard [12], the ISO/TC 307 on 'Blockchain and distributed ledger technologies'. Currently, there are over seven standards under development around the topics discussed in this book. The delivery of standards is always valuable as it enables interoperability among various platforms; however, standards are notoriously delayed.

Frequently Asked Questions

How Do I choose which blockchain is more appropriate for my project?
We have designed a specific strategy to help companies and individuals to determine what blockchain can be deployed for any project (see Chapter 7).

A large company that provides a range of services, does it need different blockchain for each service, or one blockchain can rule them all?
As each use case brings its own set of participants that would be advisable to develop a solution for each problem. Several problems are similar, and hence, solutions can be reused. It is unlikely that a consortium will enable external participants to share data.

Can we integrate the blockchain solution with existing systems?
Yes. Restructuring your existing IT network to address a problem that a blockchain can solve easily, is likely to be costly and not without risk. However, integration with a blockchain is a cost-effective route.

Is blockchain the only way to timestamp documents?
No. The problem of securely timestamping documents and information is an old one. The advent of the World Wide Web highlighted the need for secure

timestamping and there are organisations that act as a Timestamp Authority. Financial transactions over the Internet need to be reliable timestamped. The Internet Engineering Task Force (IETF) and the Internet Standards body have developed standards that have to be met by organisations wanting to be a Timestamping Authority (TSA). Anybody can use a TSA to prove the existence of a document from a point in time. The US National Institute for Standards and Technology has also produced a standard on the use of reliable timestamps [63]. The public blockchain offers an alternative way to timestamp documents. From a purely legal point of view, there is however no guarantee that a court would recognise a blockchain timestamp.

What is the difference between a public blockchain where the transactions are private, such as Monero, and a private permissioned blockchain?
Both prevent unauthorised parties from seeing the content of the ledger, but in different ways. In Monero, the cryptography enables participants only know details of transactions specific to them, although anyone can audit the ledger. In a private permissioned blockchain, only authorised participants can see all the transactions on the ledger without any restrictions.

In a private blockchain, if a participant leaves the permissioned blockchain and keeps a copy of the ledger, it's possible—though illegal—for them to leak details of transactions involving the other participants. In Monero, some participants in the transaction cannot show to third parties that that transaction occurred.

Is a permissioned blockchain less secure that a public permissionless blockchain?
Security is not an absolute concept. Security is defined with respect to an adversary. The stronger the adversary, the stronger the measure to prevent security breaches. The consensus process is the gate keeper of the security, In a permissionless blockchain, miners could be strong adversaries, and hence, the PoW is designed to stop them. In permissioned blockchains, the validators are carefully selected, and they are not assumed to be strong adversaries. Hence, the security is more relaxed.

7

Blockchain Strategy

There is always a temptation for businesses to rush headlong to adopt new technology, often driven by the fear of missing out on the 'next big thing'. Blockchain has significant potential to add value and transform the way businesses operate. Procuring the wrong technology, however, can be costly and damaging, so it's necessary to follow a clear implementation strategy. This chapter offers an end-to-end guide to enable anybody to lead a blockchain project from conception to delivery.

If Blockchain Is the Answer What's the Question?

Developing a blockchain strategy might feel challenging at first. Where should you start? What are the key features of the blockchain solution? To deliver the strategy and manage the project, we have devised a simple-follow six-step approach—see Fig. 7.1—and an associated questionnaire—see Fig. 7.2.

Step 1. Assessing Your Needs

The first and most important step is to understand the needs or the problems you would like to solve. This is harder than it looks at first.

For a start, people can be rather vocal about solutions to real or perceived problems, yet less clear about specific issues that caused the problems in the

© The Author(s) 2020
M. G. Vigliotti and H. Jones, *The Executive Guide to Blockchain*,
https://doi.org/10.1007/978-3-030-21107-3_7

Fig. 7.1 The six-step blockchain strategy

first instance. Your job is to work out root cause of the problem(s) and evaluate the impact: follow the steps below to achieve this goal.

Establish a cross-functional team Problems affect people in different areas: work out who is affected, and invite them to share their experience.

Identify at least two examples of the problem/need is required Methodically work out the details of the problem. Consider taking two concrete examples show: when, where and how a problem manifests, and more importantly how your business is affected. It can be useful to create a diagram with the basic steps outlined.

Look for a common thread in the examples It is useful to consider what key features the examples have in common, and to re-formulate the problem/issue statement in a more general way that *everyone* can agree to.

Identify risks Now it is time to think about the impact at the organisational level. What are the consequences if the problem is not addressed? What are the risks? Not all risks are equal. It is good practice to label each risk as low, high or medium depending on the financial or operational impact.

Show how the current situation is hitting the company's bottom line Ultimately, you will need to take a decision if any problem or need is worth to be addressed, by determining what is the cost of doing nothing, and how the risks impact companies' mission in short term and in the long term.

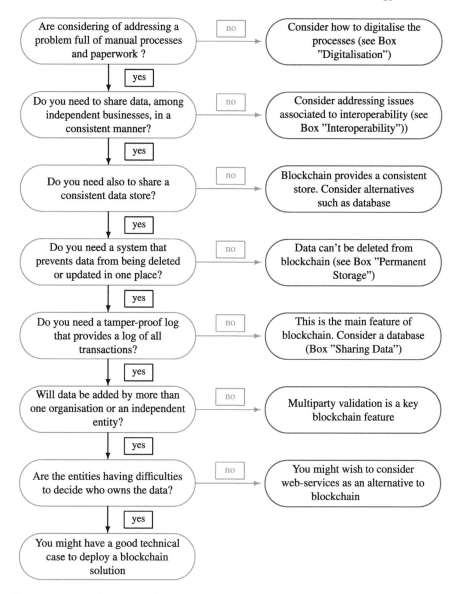

Fig. 7.2 Test to determine if you do not need a blockchain

Gathering the opinions of people impacted by a problem is important. Problems do not exist in the abstract and require collaboration to be clearly diagnosed. Organising a workshop can be a resource-effective way of completing this first step. Documenting the work carried out can be useful to gather support from the decision-makers. If, it turns out that the problem is not a priority, there is no need to follow up with the remaining five steps.

Step 2. Looking for the Solution

Once you've identified the problem, it's time to find a solution. There is the temptation to rush to discuss the technology before having worked out *your* solution, meaning the solution that works for your company. Remember that technology supports the business' goals, not the other way round. You don't need to describe the solution in a technical language, but you do need to clearly identify each single step. Delays and rising cost of software projects are often caused by the client's requests for changes, due to lack of clarity.

Finding a solution starts with addressing common issues highlighted in the initial assessment in Step 1. Follow the steps below to work out a solution to your problem:

Describe the change Start from the general solution of the need and describe the same situation, but this time, assuming that magically the problem has gone. This is a very hard step, because it requires to image something that does not exist.

Consider stakeholders Identify people that are most likely/least likely to benefit the solution. Any solution needs to be deployed by people. To successfully deliver the final project, it is important to appreciate how people are affected, or what would make the solution more palatable. Remember that a solution to some people could be a barrier for others. It is really important to carry out the research to understand the ramifications and the impact of the solution.

Risk mitigation A solution that does not, at least, mitigate the high risks is not worth pursuing. Be clear on how your solution is a tool for risk mitigation.

Measure benefits You know the cost of doing nothing, can you approximately estimate the value of the benefits of delivering the solution? One common hurdle is to accurately quantify the cost-saving, or expected increase revenues. However, remember that this is an early-stage assessment that needs to be fully revised later. Be conservative on the estimate increasing the expected costs and decreasing the revenues. It will pay off later.

It is challenging to describe a solution: with too many details, it becomes unmanageable, with too few details it is not informative. A well-defined solution requires some patience as it can be a trial-and-error exercise, and the steps above can reiterate a few times.

Digitalisation

If your industry deploys a lot of manual processes, it would benefit by automatising them. Blockchain will achieve this—but other technologies can help as well.

Let's consider the Land registry example.

Country A deploys paper land titles, which are stored in a fire protected building. When citizens need a copy of the land title, they put a physical request to the land registry office.

The problem is that a backlog of requests is very big, and often people need to visit the office twice: once for the request, once for the collection of the certificate. This leads to loss of productivity in the population.

Solution: Digital land titles accessible by citizen from a Land Registry web-page.

Interoperability

Data interoperability problems are preventing people from successfully working together.

Interoperability rules exist outside blockchain technology.

For example, data exchanged on the Internet follows several interoperability standards to enable anybody to participate.

In the case of country A's land registry, interoperability is crucial to enable solicitors and other parties to authorise registry updates when a property is bought or sold.

Permanent Storage	Sharing Data
Not all data needs to be stored for a long period. Deleting useless data is smart as it frees up space and saves energy. In some cases, data and its log (information about who created the data, when it was created) must be kept; for example, records of a court case might be needed in case on an appeal, or details of financial transactions may be needed by the regulator for investigation. A land registry is a prime example where data needs to be kept for a long period of time. A land title keeps track of all information about the land, when it was bought, who bought it, the price and rights and so on.	With blockchain, everyone can own data and prevents participants making unilateral changes. Carefully consider governance of the blockchain and the consensus process that ensures the ledger remains tamper-proof. A simple blockchain solution for the Land Registry of country A could involve a consortium of banks that issue mortgages. The banks update the land registry every time a property is sold, and a mortgage is issued. No bank owns the registry, but everyone can reliably query and update it. A possible consensus algorithm would require the appropriate government department and randomly chosen banks to co-sign the updated title. The latter will prevent collusion. If any title contains false information, then it will be known to all participants. A bank that repeatedly inserts incorrect information could be investigated.

Step 3. Finding the Technology That Works for You

As you have a clear idea of the problem and the solution, then the next step is to find out if blockchain can help. We have devised a test that will help you to clearly determine if you do not need a blockchain. Please take the test in Fig. 7.2. If you answer 'yes' to all the questions, you could be in need of a blockchain solution, and it remains to determine which one. Implementing the right blockchain plays a major role in the success of your project. This can be particularly challenging as it could involve several other parties.

If it turns out you do need a blockchain solution, then engage with a subject matter expert in the field (either from within the organisation or externally) who can help you choose the correct technology. Provide to the subject matter expert the analysis carried out in Step 1 and Step 2. You don't need to understand all the nitty-gritty technical details of the solution produced by the subject matter expert, but you can be an *informed customer* by understanding the reasons the technical solution is fit for purposes. To achieve the latter, we have selected points of consideration to evaluate the work of the subject matter expert.

How blockchain delivers the benefit outlined in Step 2 There is no point in getting a technical solution that does support the benefits you have identified—so be clear on that. Anybody proposing a technical solution must explain why and how the technology delivers your solution. In some cases, people have only a hammer that does not mean you need to become a nail.

What are the key features of the blockchain solution?
If permissioned blockchain is being proposed, then you would like to know:

- Who are the stakeholders in the blockchain consortium?
- Are you joining an existing consortium, or are you going to create your own consortium?
- Governance framework—structured agreement on running the consortium
- Which participants validate the transactions?
- Who can read and query the ledger?
- What is consensus process and why is fit for purpose?
- What is the format of the data stored on the ledger (cryptographic-hash, encrypted data, plain-text data)?
- What are on/off-chain data requirements (what stays on blockchain, what data is stored somewhere else)?
- How the performance of the solution can impact the business?
- Is the delivery of the solution about customising an existing platform?

If permissionless blockchain is being proposed, then you would like to know:

- What is consensus process and why is fit for purpose?
- What are chain/off-chain data requirements?

- How is the data integrity guaranteed?
- Is there any personal data that needs to be carefully considered?
- Is there already an existing (open source) solution that can be easily deployed?
- What are the technological risks and mitigations?
- Determine the Return on Investment (ROI).

Step 4. Prototype

The first three steps are part of the planning, and the rest of the steps are about ensuring a successful delivery. The purpose of the prototype is to determine how a solution works with a small budget and minimal risks. A prototype is not the complete product, but merely a version that contains the important features. The most difficult decisions relate to what features of the solution will be left out. The key features and the timeline to develop a prototype should be agreed in advance and with *Agile software development*[1] softeware development can last from a couple of weeks to a couple of months, depending on the complexity of the prototype, and the size of the team and wether you are customising an existing (possibly open source) solution or starting from scratch. Meeting the delivery deadline is a must, do not drag project beyond necessary. Ensure that everybody is clear about what features must be part of the prototype, and what features will be left out and why can be added later.

Ideally, the prototype should be modular meaning additional features, for the full solution, can be added later without changing the overall design or starting the work again.

Key points to remember about the prototype:

1. Ensure the user interface is as smooth as possible; this will help in the validation process.
2. Only implement the relevant functionality, not necessarily the basic one.

---

[1]Agile is a software development iterative approach where the team delivers work in small, time-limited increments. Requirements and software are continuously evaluated to enable teams to quickly respond to changes.

3. If you are operating in a consortium, ensure that a representative of each stakeholder is fully engaged in this phase.
4. Create a culture of collaboration where issues about the software development can be discussed and resolved quickly.

Remember, the risk involved in doing nothing is greater than the risk of doing the wrong thing.

Step 5. Validation

Validation allows you to make sure that objectives and benefits outlined in Step 2 can be delivered. The validation is divided into two parts:

Testing the prototype The end-users must test the prototype and provide feedback. If you are working in a consortium, make sure that every stakeholder performs the user's tests. Devise a series of tests on the features of the prototype, with clear, easy-to-follow step-by-step instructions; for each task, don't forget to include a well-structured questionnaire to assess also subjective features such usability of the user interface. Finally, discuss with all stakeholders (of the consortium) the impact of the solution on their business and collect the feedback about the prototype.

Assessing the solution After having collected the responses from the stakeholders, get the report of the developers tests to find out any issues found, for example bugs, performance that needs to be addressed moving forward. It's time to evaluate if the technical solution is really delivering the benefits identified in Step 2.

The prototype phase, with the feedback both the stakeholders and end-users should provide an insight about the challenges in delivering the full solution. The key questions to ask are:

1. Has the prototype been delivered on time and within budget?
2. Has the overall cost of the project changed as a result of the work on prototype?
3. What are the technical problems emerged during the prototype phase?
4. What have you learned about the solution with prototype?
5. What are the challenges faced by from the end-users?
6. What features have been positively received from the end-user?

Next it is important to understand if the project is going to deliver what everyone has signed for. The key questions to ask are:

1. Do the users-tests indicate that the prototype is well received and addresses the original problem?
2. Are all stakeholders engaged with the current solution?
3. Do stakeholders understand how the blockchain will impact their way of operating?
4. Given the current status of the project will both the benefits (including the ROI) be delivered?
5. Are there any new obstacles or risks? How are they going to be managed?
6. What exactly needs to change to deliver the benefits of the original solution?

Answering these questions could require to revisit some parts in Steps 1, 2, 3 or 4, and to work out a slightly different technical solution. At this stage, clarity about what works and what does not work ensures that the project will be successfully delivered.

Step 6. Scale

The final stage is to decide how to scale the initial prototype based on the validation. Typically, this means proceeding in small steps and performing incremental micro-evaluations. If a permissioned blockchain is deployed, the governance plans need to be in place sooner rather than later. It is easy to achieve 80% of the project in 20% of the time, but it is very frustrating to wait for the 80% of time to deliver the remaining 20% of the project. Agile development enables incremental deliveries within short timeframes that help to manage changes and frustration.

For any software project is critical to never underestimate issues related to cybersecurity: in our experience, small details can make a huge difference in this field. Security testing by an independent party should be part of the scale plan.

A successful prototype will allow you to build upon it, confident that benefits and risks have been quantified, the technology will deliver value, and everyone in the consortium is engaged.

You can safely continue the journey towards innovation.

Going Alone

Does this mean that to deploy blockchain a consortium is always needed? Not, quite.

It is possible to meaningfully deploy blockchain without the need of a consortium.

Accepting cryptocurrencies payment In the Western world, payment systems via bank transfer or credit card are very good, while only 11% of the cryptocurrency transactions are payments[2] [99]. Technological companies like Microsoft and Wikipedia accept bitcoin to increase the customer base by offering an alternative way of paying. If you do not know where to start, consider engaging with a payment processor, who would convert the cryptocurrency into fiat currency on your behalf, and take the responsibility for the compliance to relevant anti-money laundering and know-your-customer regulation. From an operational point of view, this is not more complicated than dealing with other payment processors.

Certifying and timestamping documents In creative industries, for example, people need to show that a document existed at a specific point in time. It could be to show the original certificate of a piece of art to prevent forgeries, to enforce payment of copyright fees or to prevent to steal other people's ideas. Some companies may need to securely timestamp documents as part of the compliance process, for example, to show that a doctor was appropriately trained at a time of medical mishap, or that very stringent procedures were in place when at time of a safety accident. Permissionless blockchain is a reliable way to occasionally timestamp documents—see Chapter 6—for more information.

Summary

Blockchain has moved from being the next big thing to become a more mature technology that is attracting interest and take-up across a number of sectors. To deploy blockchain technology in a plan is needed. If you like to use the six-step strategy described in this chapter, you can download the template from the associated website https://www.sandblocksconsulting.co.uk/.

---

[2]A market size is approximately $3.7 billion at the capitalisation at the time of writing.

8

Smart Contracts

There is a buzz of excitement surrounding smart contracts, with considerable speculation about the role they will play in changing the face of e-commerce. To back this up, there has been significant investment in smart contract technology from small organisations as well as much larger consortia.

Some research suggests that smart contracts will become more popular, and by 2022, more than 25% of global organisations will use blockchain [23]. That's a big claim for something that most people have never heard of, including, in all probability, a majority of the world's lawyers.

It is too early to say whether or not this is all hyperbole; however, smart contracts do feature prominently in blockchain applications, so it is important to understand what they are and how they work.

How Smart Is a Smart Contract?

In recent years, the word 'smart' has increasingly been used to describe digitally enhanced objects. 'Smartphone' is the most obvious example, but there are many others including 'smart city', 'smart bulb', 'smart refrigerator', 'smart camera' and so on. What links all of these developments is the technological advancements they deliver. With this in mind, it would be logical to assume that a smart contract is simply a digitally enhanced form of a conventional contract. While that is partly correct, it is only part of the story. In this chapter, we will look at the many different uses of smart contracts and the impact they are having on business.

© The Author(s) 2020
M. G. Vigliotti and H. Jones, *The Executive Guide to Blockchain*,
https://doi.org/10.1007/978-3-030-21107-3_8

It's important to understand that smart contracts are actually agreements, written in computer code, a piece of software that executes or enforces some of the terms of the agreement. Using the word agreement here, rather than contract, is deliberate. A contract is a legally binding agreement, which can be enforced in a court of law. An agreement, however, is a stipulation of clauses which sets out obligations without being legally binding. All contracts are agreements; not all agreements are contracts. Some smart contracts can be legally binding, but others will need to be tested in court. In some extreme cases, there could even be a need for new legislation.

What Is a Contract?

Under the laws of England and Wales, for a contract there are four key elements:

- Offer
- Acceptance
- Consideration and
- Intent to create legal relations.

Most contracts are written because then it is easier to show evidence of the original agreements. The problem with purely oral contracts is that they are difficult to enforce.

As the law doesn't always specify the means to set out a contract, it could be claimed that a piece of code that satisfies the four conditions above is legally binding. One of the challenges is for the code to provide suitable evidence of the four elements of the contract, hence we can envisage a distant future where solicitors, barristers and judges would need to learn how to read and interpret code, and to correctly run a piece of software.

Myth Busting and the Role of Smart Contracts

A lot of myths have developed about smart contracts, which, in the last ten years, are often viewed as a by-product of blockchain technology. Myth number one, which requires immediate dismantling, is that they are new. In fact, many of us already use smart contracts in our everyday lives. For example, making a 'contactless' payment directly with a bank card for a tube or bus journey and hiring a city-centre bike are all examples of deployment of smart contracts.

On the blockchain, a smart contract is just a piece of code. The difference is that a contract is code that located on the blockchain that is recorded on the ledger. All participants in using the contract will:

- Have a copy of the code.
- Be able to audit the contract.
- Interact with the contract to provide input.

Once registered, a contract on the blockchain cannot be modified.

Traditionally, bike hire in London would involve purchasing a ticket or a subscription for a set amount of time. Hire terms on the ticket (or some other document), would be proof of a contractual relationship. However, when we use a credit card to hire 'smart bikes' from docking stations located in a 'smart city' like London, there are no terms to sign-up to. The bike is released with a simple tap of a debit or credit card and the correct amount of money is debited from our bank account when the bike has been returned. Everything is managed automatically without any human intervention.

There's a strong argument to be made for this type of transaction, not least because cutting out human involvement speeds up the process and makes it much less convoluted. Furthermore, digital transactions are simpler to log and archive, and as a result easier to audit.

To summarise, the benefits of using smart or automated contracts in this example are:

- Improved performance
- Cost reductions
- Transparency.

The advent of the Internet has increased the deployment of this kind of 'smart agreements' so much so that Article 9 of the European Union's Electronic Commerce Directive [27] requires member states to ensure that their legal systems don't create barriers for electronic contracts or result in them being deprived of their legal effectiveness.

Although the directive uses the term 'electronic contract', it essentially covers smart contracts which can be run automatically. Under the directive, our example of the bicycle hire contract would be legally binding in English law (at the time of writing the UK is still a member-state), as it satisfies all the conditions of a contract.

A Brief History of Smart Contracts

So, if smart contracts aren't new, why are they generating so much hype? To find an answer, we need to look back to the time when the first smart contracts were created. An American computer scientist, Nick Szabo, is thought to have first used the term smart contract in an article in 1996 (Szabo, incidentally, is rumoured to be 'Satoshi Nakamoto', the mysterious creator of bitcoin, although he strongly denies this). He wrote [104]:

> A smart contract is a computerised transaction protocol that executes the terms of a contract. The general objectives of smart contract design are to satisfy common contractual conditions (such as payment terms, liens, confidentiality and even enforcement), minimise exceptions both malicious and accidental, and minimise the need for trusted intermediaries. Related economic goals include lowering fraud loss, arbitration and enforcement costs, and other transaction costs.

While this definition is somewhat interesting, it's also very mechanical.

Two years later Szabo observed: 'Whether enforced by a government, or otherwise, the contract is the basic building block of a free market economy'. Szabo was convinced that a new digital era would change people's lives, an accurate prediction by any measure. He also foresaw how the integrity of contractual relationships could be preserved, writing [105]:

> New institutions, and new ways to formalise the relationships that make up these institutions, are now made possible by the digital revolution. I call these new contracts 'smart', because they are far more functional than their inanimate paper-based ancestors. No use of artificial intelligence is implied. A smart contract is a set of promises, specified in digital form, including protocols within which the parties perform on these promises.

How Should a Smart Contract Be?

Verifiable Because smart contracts are written in code, it is important to verify that the code contains agreed clauses. Verification also helps determine whether there has been a breach of any clause in the contract.

Observable Any complex contract contains a dependency of obligations, meaning that certain obligations arise only if specific requirements are met. Parties to a smart contract must be able to observe how obligations are met, so that early breaches can be detected. For example, bicycle hire could become cheaper after two hours of continuous hire. As a party in this smart contract, it should be made clear to the person hiring the bicycle at what point the time starts (after payment, or when the bike is removed from the docking station?), which makes it possible to determine that the obligations have been respected.

Enforceable Ideally, there should be little or no need for an external party to ensure that the obbligatations of a contract are met; for example in the case of the bicycle hire, deduction of the correct amount to pay for the hire should happen automatically, without human intervention. Also, it should not be impossible for the bicycle hire company to withdraw the money for the rental of the bike. Such a possibility would potentially require court action as redress.

Privity Privity is described by Szabo as 'principle that knowledge and control over the contents and performance of a contract should be distributed among parties only as much as is necessary for the performance of that contract' [105]. He also noticed that cybersecurity principles and fit-for-purpose software should be used to ensure smart contract obligations are not inadvertently shared with other parties, which 'roughly corresponds to the goal of privity'.

Szabo realised that contracts are essential for trust in functioning societies. Since he made his predictions in the 1990s, computer scientists and mathematicians have created the technical tools to automate contracts. In fact, some early technology that enables smart or automated contracts already existed at the time when Szabo was sharing his thoughts with the world, including Digit cash by Chaum [72], a payment system that protected user privacy. Today, as smart contracts develop and replace some traditional contracts, they have the potential to significantly reduce costs and speed up execution.

Szabo was also quick to understand the fundamental difference between contracts written in English (or other natural languages) and executed by humans, and the more opaque smart contracts, where clauses and contract execution are not evident. He explained that to make them a valuable tool for society, they needed to be: verifiable; observable; enforceable; and have privity (see Box '*How Should a Smart Contract Be?*'). Szabo also forecast that in time it would be possible to buy and sell houses without the intervention of a solicitor, and intellectual rights could be distributed automatically. Smart contracts or smart agreements would be part of the fabric of society, provided they satisfied the four principles shown above. The contracts would, in his opinion, lower legal barriers, slash transaction costs, cut the time taken to execute a contract and provide an opportunity to create new types of business.

Where Does the Blockchain Fit In?

Of course, Szabo was forecasting a bright future for smart contracts at the dawn of the Internet age, long before the development of the blockchain. Today, the term 'smart contract for blockchain', which came about with the development of Ethereum blockchain (see Chapter 5), is a familiar one for technophiles. The Ethereum platform extends Bitcoin by enabling arbitrary code, known as 'contracts'.[1] Some of these contracts have enabled activities that are more typically associated with the legal profession, such as contracts used by companies to raise funds (see sections 'Real Life Smart Contracts' or 'Create New Cryptocurrencies or Tokens').

This isn't the only reason for the increasing popularity of smart contracts. As we've already seen, one of the main functions of a blockchain is to create a chain of trust among participants. Smart contracts on the blockchain provide a clear illustration of how this can be achieved by allowing all participants to verify and enforce clauses on the ledger and watch its progression.

For further evidence of this, let's return to our example of bicycle rental. Someone renting a bicycle doesn't need to audit or even see the software that releases the bike and collects the payment. They assume that everything will

---

[1]The programming language for Ethereum is called Solidity and is classified in computer science terms as 'object oriented'. In other object-oriented programming languages, the common terminology is 'class', instead of contract.

run smoothly. However, if they are overcharged, the hire company has a clear and obvious obligation to repay what is owed. Legally, this is because the software forms part of an unwritten contract; consumers have acquired rights under English or EU law, regardless of how the terms of the contract are implemented; the payment software is simply there to speed up an otherwise laborious process. This highlights how smart contracts can be part of a bigger legal contract framework, where some clauses are automated. This is sometimes called *smart contract code* [75].

What Is a Smart Contract on the Blockchain?

A smart contract on the blockchain is a piece of code that represents the terms of an agreement among parties. The obligations are enforced via the consensus process when the parties deploy the contract. A smart contract on the blockchain enables participants to:

1. Inspect the code to ensure it meets the agreed clauses.
2. Be reassured that an agreed contract, once registered on the blockchain, is tamper-proof.
3. Be certain the contract executes in the same way for all participants.

In the case of complex, multi-party contracts, it would be very difficult for one party to run a version of a smart contract on behalf of everyone. Ordinarily, each party might require a solicitor to read the contract and provide reassurance about the clauses and the implication of the contract. It's the same with a smart contract. The blockchain enables all parties or participants, located anywhere in the world, to automatically oversee the contract, without the need of a third party. We can see this with the Ethereum blockchain, which allows participants to inspect and audit the clauses.[2]

Auditing a smart contract is a job for a competent software engineer who will ensure that cybersecurity breaches are minimised, and that the agreed clauses are in the code. Once recorded on the blockchain, the code cannot be unilaterally modified.

---

[2]Test this for yourself by looking at the code written by one of the funders of Ethereum at https://etherscan. io/address/0x972a2da1f9d1dc0b01d313e52ffe916bb5e9a2c1#contracts.

Smart contracts can run on both public and permissioned blockchains. Public blockchains, such as Ethereum, are particularly suitable because they:

- Facilitate digital payments (cryptocurrency).
- Create trust among participants thanks to their robust consensus mechanism.

The table below summarises the difference between a standard and a smart contract:

	Traditional contracts	Smart contracts
Offer and acceptance	Natural language	Code
Consent	Execution of signature	Execution of electronic or digital signatures
Consideration/Payment	As agreed	Built in
Escrow	Trusted third party	Built in
Audit/Second opinion	Competent solicitor	Competent software engineer

AXA's Flight Delay Compensation Insurance

In September 2017, the multinational insurance giant AXA introduced a delay repayment scheme covering 80% of flights worldwide, including all EU countries. Crucially, the scheme uses a smart contract, which makes it easier for passengers to claim for delayed flights, while at the same time minimising the risk of fraudulent claims.

How Does It Work?

When a flight is delayed beyond an agreed time limit, a payout is made within hours without the claimant having to do anything. It's a win-win: the claimant doesn't need to waste time filling out endless forms and AXA is protected from bogus claims as only a subscriber to the scheme with a valid claim will be compensated (Fig. 8.1).

AXA's Smart Insurance Scheme

A new insurance, which is recorded using a smart contract code, can only be added if:

- A flight is identified.
- The arrival time is clear.
- The premium and the indemnity are both agreed.

The smart contract updates the current status of the ledger with the new parameters given by the user. Registration via a user-friendly web-interface means new subscribers don't need to interact with the Ethereum blockchain.

After the insurance created, the customer has right to withdraw from the insurance. AXA doesn't use the Ether cryptocurrency, as might be expected, for premium collections or payouts. This is probably because most subscribers would be unfamiliar with cryptocurrencies.

Payments are made using another piece of software that is integrated with the blockchain. The code is not part of the blockchain and crucially is not available for inspection. The contract is legally binding, but not because it is smart! When a user registers with AXA for this product, they agree to terms and conditions that form the contractual obligation, well beyond what is included in the smart contract.

To sign-up to the scheme, travellers submit their details at https://fizzy.axa; if they are happy with the quote they receive, they subscribe and pay a premium. A third party data provider has access to the smart contract, and updates the records on the blockchain when a flight is delayed. AXA can chose to comply with its contractual obligation based on the evidence provided by the blockchain.

This smart contract, which runs on the Ethereum platform (see Chapter 5) and is similar to a traditional contract, can easily be inspected. Indeed, if you understand Solidity, the programming language used by the Ethereum blockchain, the programme will feel very familiar. It can be found at https://etherscan.io/address/0xe083515d1541f2a9fd0ca03f189f5d321c73b872.

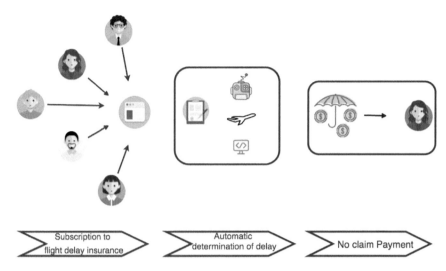

Fig. 8.1 Graphical representation of the steps involved in the Axa Smart Contract

Company Funding Using Smart Contracts

Companies can use public blockchains to raise capital in exchange for tokens through *Initial Coin Offerings*, also known as 'cryptocurrency crowd sales'. Companies start by issuing a 'white paper', a type of customised business plan, informing potential investors about the merits and the risks of the project. Tokens are designed to represent something, a stake in the enterprise or access to a service or a receipt for a donation. The white paper specifies whether there is a minimum funding cap to allow the project to proceed. Typically, but not always, funds are collected in Bitcoin, Ether or other cryptocurrencies, and in some cases in fiat money as well.

The next stage is the release of the tokens, usually after due diligence and money laundering checks have been completed (funds are returned to investors who fail these checks). Coins/tokens are then listed on cryptocurrency exchanges and traded against fiat and other cryptocurrencies. Their value depends on the success or otherwise of the project and the attractiveness of holding the tokens. This process is similar to an Initial Public Offering (IPO), or private capital fundraising by start-ups. In the case of an ICO token, distribution is carried out by one or more smart contracts. It is the smart contract which computes how many tokens an investor should get, how much funding has been raised and the legitimate period of sale. Funds are rejected when sent at the wrong time and tokens are distributed automatically after funding. Depending on the terms set out in the white paper, the smart contract can

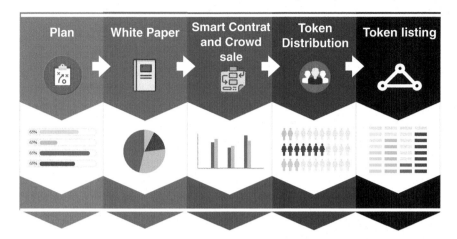

Fig. 8.2 Graphical representation of the steps involved in an ICO

be used to lock the tokens for a vesting period to prevent investors who have bought at a discount triggering a massive sell-off (Fig. 8.2).

This is just one way in which an ICO differs from a conventional IPO. In the latter, an investor puts their trust in the process, including regulation and the solicitors handling payment and distribution of shares. By contrast, an ICO investor accepts that the agreed clauses and due diligence are performed by the smart contract.

At this stage, readers may be asking themselves whether or not this type of smart contract is, in fact, legal. What would be the outcome, for example, if an investor didn't receive tokens they had paid for because of a mistake in the smart contract? Would they have legal redress and where should the redress be determined? The legality of ICOs and of smart contracts depends on a number of factors including the rights associated with the tokens, and in which jurisdiction the sale and marketing of the tokens were made. Different countries have different laws, which impact on companies and investors alike (see Chapter 9).

The Smart Contract of The DAO

Ethereum blockchain split into two cryptocurrencies, 'Ethereum' and 'Ethereum Classic' (see Chapter 5), following the hacking of the Decentralised Autonomous Organisation (The DAO). It is worth taking a little time to look at the role played by the smart contract (see https://etherscan.io/address/io/address/0xbb9bc244d798123fde783fcc1c72d3bb8c189413) in the dramatic split.

The DAO was established to enable a group of organisations or entities to operate like a venture capital fund, but instead of having people manage the fund, a smart contract was effectively in charge, removing the need for a central governing authority [88]. The DAO also allowed for democratic decision-making when it came to spending the funds. Initially, funds were from investors providing Ether in exchange for The DAO tokens. When the smart contract was implemented in May 2016, one Ether bought about 100 DAO tokens, though the price was not fixed and changed over time. Early investors received a discount and DAO tokens provided the right to vote on what projects would receive funding. The DAO raised 12.7 million Ether, which was equal to more than $150 million in 2016 making it the biggest crowdfunding project of its time.

The highly ambitious project eventually failed, but the goal of automating governance continues to be an appealing concept.

Interestingly, in the USA, the Security and Exchange Commission (SEC), which oversees and regulates the distribution of securities, concluded that The DAO crowd sale amounted to a sale of securities to investors. If security instruments are marketed to US citizens, any company, regardless of where it is incorporated, needs to be either registered with the SEC or ask for an exemption. The DAO failed to comply and in an announcement on its website [33] on July 2017, the SEC concluded that Slock.it,[3] the company behind the project, had breached US Federal Laws. The letter was clear warning for other companies that wanted to pursue similar activities. The SEC decided not to prosecute Slock.it on this occasion.

The Legal Side

Looking back at our examples, the legality of a smart contract on the blockchain is partially defined by national legal frameworks.

AXA operates in Europe and offers flight insurance to consumers via its website. Deploying a smart contract on Ethereum is simply a technological device that doesn't change agreed terms and conditions or a consumer's fundamental rights. The legal status of the crowdsale smart contract changes from country to country (see Chapter 9).

Smart contracts can run on both public permissionless blockchain and permissioned blockchain. Both carry different risks. On the former, legal disputes can be problematic because users can only be identified by their address. In many jurisdictions, a contract is legally binding whenever the parties are

---

[3] See https://slock.it.

human or legal entities. Furthermore, in some cases, there is a need for parties to know who the other parties are in the contract.

Because of this, the legal challenge inherent in a public permissionless blockchain is to meet the commercial need for contract formation when identification is limited to a blockchain public address. There is no central authority that can unlock or help to circumnavigate this issue unlike, for instance, mobile phone operators who can be legally compelled to reveal a user's details. On the other hand, smart contracts that run on permissioned blockchains reduce legal risks because participants are pre-selected, and their identities known. Typically, they have already agreed to a set of governance rules, as we saw in Chapter 6. This provides an opportunity to create a legal framework that takes account of potential outcomes of deploying smart contracts, including the jurisdictions to settle disputes. In other words, smart contracts would be treated as any other technology.

As the philosophy of the blockchain, even a permissioned one, is to create a chain of trust, the idea of legal action is a long way from the founding principle. Today a system of arbitration is being considered, allowing disputes to be resolved within the blockchain community. It will be interesting to see how successful these initiatives are in the long term.

Of course, if you are considering deploying a smart contract on the blockchain or being party in a smart contract solution, it's important to balance the legal risk against the benefits, which are:

Cost reduction and speed of operations Smart contracts reduce the hours spent on traditional business processes and improve performance.
Reduction of human error This should be counterbalanced by the fact that code can contain bugs, and can deliver errors that are often can be costly and more difficult to spot.
Enable real-time data updates Automated data updates could be more accurate because it reduces the risk of human error.
A degree of disintermediation A smart contract cuts the need for some trusted third parties.

These benefits have led to relevant stakeholders developing use cases for smart contracts in banking and insurance.

When Automation Isn't the Answer

So far, we have looked at the benefits of using smart contracts on permissioned and permissionless blockchain and some readers may now be wondering if it is possible to make all contracts 'smart'.

Clearly defined operational clauses can be easily automated. However, complex contracts are full of so-called non-operational clauses or legal phrases that are only understandable to lawyers. These clauses, with phrases such as 'best endeavours', 'reasonable endeavours' or 'good faith', are a barrier to automation. The phrases refer to intentions of the contractual parties and involve judgement that often cannot be postulated in advance or captured in code.

Take the phrase 'a senior manager needs to take reasonable steps' to ensure a financial institution is compliant with the FCA regulation. It is nearly impossible to define in advance an exhaustive list of 'reasonable steps' for any manager to take. However, a judge in a court would be able to form an opinion on the conduct of a person and determine if that conduct was in fact equivalent to taking 'reasonable steps'.

To sum up, despite the increase in the number of smart contracts, currently there is a relatively small subset of contracts that can be automated, as contracts that do not involve operational clauses are very difficult to be put into a smart contract. It is possible that wide deployment of smart contracts will lead to changes in the legal profession: procedural and repetitive tasks will probably be automated in the future, leaving the legal profession to carry on interpreting the principles of the law.

Smart Contracts in the Financial Sector

Banks and other financial institutions have a record of being quick to adopt new technology. Algorithmic trading, when an algorithm buys or sells shares, is an example of a smart contract.

As a compliance sector, finance frequently deploys contracts; delivering smart legal contracts is an obvious next stage for the sector. This means handling a legal framework where certain obligations may be put in code. In particular, several of the financial instruments such as options, futures and swaps and bonds, whose terms require mathematical computation, are already run by computers/machines, yet the contracts that deal with such instruments are still largely formulated in a traditional way. It seems natural to formally link the contract with the piece of code that performs the computation.

The International Swaps and Derivatives Association (ISDA) is one organisation that has been working to make contracts more manageable. In the last two decades, it has created the Master Agreement Template (MAT), a contract template that typically works for 'over the counter' (OTC)[4] financial contracts [77]. The terms of complex financial products can be quite lengthy and require a lot of negotiating. The MAT enables institutions to agree on the broad terms of the contract, and prevents them re-signing the same terms when a new financial instrument needs to be traded. Once the MAT has been signed, the documentation of future transactions between parties is reduced to a brief confirmation of the specific terms of the transaction, delivery dates, interest rates and so on. When a transaction is entered into, the terms of the master agreement do not need to be renegotiated.

Recently, the ISDA has published a number of white papers [34, 47] to develop the MAT towards the smart derivative contracts, which contain clauses that can be automatically performed by code. Terms that can't be automatically performed are expressed in natural language. These legally binding contracts are speeding up execution and reducing the costs of creating them.

It is possible that elsewhere in the financial sector, and in other sectors that are heavily reliant on compliance, this model will be followed as it reduces the legal risks and maintains the benefits of automation of smart contracts.

How Will Smart Contracts Affect My Business?

While it is possible—perhaps even probable—that smart contracts will become increasingly popular in business, any company considering deploying a smart contract on the blockchain would be well advised to obtain a technical, regulatory and legal assessment before proceeding. Companies frequently underestimate the risk of software bugs and most software contains bugs. Being able to clearly assess technical risks and having plans in place to combat them are key in the deployment of smart contracts. If smart contracts do become mainstream, their deployment will be standardised, with a range of commonly defined terms. In this way, different companies will be able to understand and subscribe to them. The International Standard Organisation (ISO), which is responsible for the creations of the world standards, has created

---

[4]OTC contracts are privately negotiated and traded, meaning that the parties decide their terms without going through an exchange (or other intermediary). Financial products, such as swaps, forward rate agreements, exotic option and so on, are almost always traded in this way.

a working group dedicated to standardising blockchain and DLT technology, including smart contracts. Development of standards is notoriously slow, but once completed, this will help with the deployment of new smart contracts.

Summary

Smart contracts on the blockchain are already with us and, as we have seen in this chapter, their popularity is increasing.

Of course, a contract's validity depends on the legitimacy of its formation and the execution of its clauses. But a contract's enforceability also depends on national jurisdiction. For example, under English law, a simple email can result in the creation of a legal contract. This principle could also apply to smart contracts.

A possible near-term application of a smart contract is for the legal contract to remain in natural legal language, but for certain actions to be automated via a smart contract.

There is a difference between smart contract code, which refers to code that is designed to execute certain tasks, and a smart legal contract, which refers to elements of a legal contract being represented and executed by software. Certain operational clauses within legal contracts lend themselves to being automated. Other non-operational clauses—for instance, the governing law of a contract—are less susceptible to automatisation.

Frequently Asked Questions

Do I need to read or run code when using a smart contract?
It's possible to interact with a smart contract in many ways. A webpage is one way to interact in a user-friendly manner. Using an Ethereum wallet is another.

Why is the word 'contract' used to describe something that, essentially, is a piece of code?
One reason: convention!

Can smart contracts be deployed on any blockchain?
In principle, yes. However, in practice, we are not there yet. Some blockchains share similar technology such as Ethereum, Ethereum Classic or Tron.

How can I minimise the risk posed by bugs in smart contract software?
It is always advisable to have the code audited by a specialist software company.

Will Artificial Intelligence (AI) result in more non-operational clauses being translated into code?
Translating legal reasoning into code has been the subject of academic research in the last 20 years. There are examples where AI and machine learning techniques have been successfully used. One example is medicine, where diagnosis can be delivered by a clearly defined set of measurable parameters. In contrast, automating the role of a GP isn't likely to happen in the foreseeable future.

Will solicitors need to learn to code?
It is possible, particularly if automation of contract clauses becomes commonplace that they will learn some 'higher-level languages', meaning a language 'closer' to the requirements. It's likely that a solicitor will want to control the code-writing process to ensure the original business meaning isn't changed.

Is the consensus on a blockchain a smart contract?
Not everything that can be automated is a smart contract. If that was the case, any computer program could be defined as a smart contract. A smart contract is a piece of code that embeds some form of agreement among parties that could potentially be elevated to the rank of contract.

9

Regulation

At its best, technological innovation benefits business and consumers, by cutting costs, increasing profits and creating new jobs. However, it can also have unforeseen risks, face resistance from technophobes and expose businesses and consumers to fraud. It's the job of regulators, therefore, to protect consumers and markets without stifling innovation.

This chapter looks at how some governments and regulators are responding to the growth of cryptocurrencies and the blockchain and what action might be required to ensure the benefits of both outweigh potential disadvantages.

Why Regulation Is Needed

As an increasing number of people are getting involved with cryptocurrencies, regulators have started to take a more proactive role in the crypto-industry. The development of various new activities such as Initial Coin Offerings (ICOs) and new industries such as crypto-exchanges has brought varied responses—from offering guidance on how to navigate this new brave world, to banning services and delivering new legislation. Regulation changes from jurisdiction to jurisdiction, which makes it difficult to discuss regulation in detail; however, it is possible to discuss the challenges and the principles of regulation.

Let's start by looking at why regulation is needed. Financial regulators have the crucial role of upholding the integrity of markets by, for example, encouraging fair competition, ensuring financial inclusion and protecting consumer's rights. Other regulators, including government agencies, clamp down

© The Author(s) 2020
M. G. Vigliotti and H. Jones, *The Executive Guide to Blockchain*,
https://doi.org/10.1007/978-3-030-21107-3_9

on illegal activity, such as tax avoidance, money laundering and finance of terrorism. Both types of regulators are keeping a close watch on (ICOs), cryptocurrencies and blockchain technology, which have the potential to benefit people around the world but could also be hijacked by unscrupulous criminals. Most regulators—and some central banks—have already issued strong warnings about the risk of investing in cryptocurrencies, due to their volatility and protections for consumers in case something goes astray.

Some countries, including the EU, Australia and Canada, have extended existing laws; others, including Bangladesh, Thailand and China, are preventing their own financial institutions from facilitating cryptocurrency transactions. A third group, which includes countries in South America and Asia, has banned cryptocurrencies altogether—see Fig. 2.5. Despite this, investor enthusiasm is growing for both blockchain technology and cryptocurrencies. Spain, Belarus, the Cayman Islands, Luxemburg, Switzerland, Malta and Gibraltar are among the countries that have recognised this reality and have issued specific legislation aimed at providing clarity and attracting new businesses.

Crypto Money Laundering

According to the United Nations Office on Drugs and Crime, between 2 and 5% of global GDP (i.e. $800 billion and $2 trillion) is estimated to be laundered annually [40]. Europol, Europe's police agency estimates $4.2–5.6 billion, equivalent to 3–4% of the continent's annual criminal takings, are crypto-laundered. Cryptocurrencies are mostly used for legitimate activities, yet it cannot be denied that they can be a new means to help criminals. While the reality is that crypto-laundering is still a small share, the anonymity of transactions and global availability of cryptocurrencies facilitate transfer of money by bypassing global regulation.

How does crypto-laundering happen? Let's consider the Colombian cocaine cartel: criminals deploy crypto-exchanges to convert dollars into anonymous cryptocurrencies, held in a wallet and swapped into pesos on another exchange. The pesos are then withdrawn in cash, spread over in sums small enough to avoid suspicion to the local banks.

Regulation of blockchain and cryptocurrencies is evolving rapidly, and areas where the regulators are focusing are:

The status of cryptocurrencies If cryptocurrencies or cryptocoins are not legal tender or fiat money, as discussed in Chapter 2, what are they? Are they recognised in some way or other by governments?

Initial Coin Offerings and cryptoassets Public funding activity falls under the umbrella of financial regulation in several countries. How does the presence of tokens affect the current regulation?

Taxation It seems self-evident that tax evasion through cryptocurrencies should be prohibited. But what about tax avoidance? When people or businesses make earnings out of cryptocurrencies, should these be taxed? If cryptoassets are exchanged for goods or services, should the taxman collect the appropriate fee?

Exchanges and trading As discussed in Chapter 4, people can directly transfer cryptocurrencies. However, an ecosystem of operators and providers has grown. Broadly, we identify:

- Processing services
- Wallet providers
- Exchanges
- Trading platforms
- Other actors, like merchants, payment facilitators, ATM.

If individuals have access to these services, how can the regulator protect their consumer rights?

The technology itself If the technology provides specific benefits, should it ensure buyers are getting exactly what they paid for?

ICOs

An ICO—also known as crypto crowdfunding—is a means of raising finance by asking a large number of people to make a small contribution (typically in the form of cryptocurrencies), in return for newly minted tokens (see Chapters 5 and 8).

Name	Amount (million)	Time
Tezos	$232	2017
Filecoin	$257	2018
SIRIN Labs	$159	2017
Bancor	$153	2017
Status	$108	2018
QASH	$105	2018

After they have been issued, tokens are usually, but not always, registered and traded in one or more crypto-exchanges (see Chapter 5). The total amount raised by ICOs between 2014 and 2018 was around $12.4 billion (see Fig. 2.6). ICOs can reach a similar level of financing as Initial Public Offerings; a few major players have raised in excess of $100 million, as the table shows.

The Mystery of QuadrigaCX

The CEO of QuadrigaCX, a Canadian crypto-exchange, died due to health complications in December 2018 while travelling to India. He was the sole person in the company to possess the private keys of customers' cryptocurrency accounts worth $143 million. Because nobody in or outside the company can retrieve the private keys, now customers' funds are locked; QuadrigaCX had no other choice but to file for bankruptcy on January 31, 2019.

This is an area where regulation can have a positive impact to address the new challenges posed by cryptocurrencies. To protect customers, in ordinary financial businesses, it is required that customers' funds are kept in custody, segregated from business activities, and available to satisfy customers' transfer requests. In the crypto world, the concept of segregation does not make much sense. Funds are kept in a digital wallet that can be stored anywhere. What provides access to the crypto-funds are the private keys. The notion of custody has moved from defining a process to finding a technical solution, that not only protects the customers, but eliminates risk.

A positive regulation should define clear rules for custody of private keys to protect customers funds from internal threats and hackers, and at the same time, enable several individuals to have access to the keys, in specific circumstances. Key management is a well developed area of cryptography, and the unfortunate consequences of the demise of QuadrigaCX could have been adverted.

Participants in ICOs, or people who purchase tokens in secondary markets, are attracted by the prospect of handsome returns when selling once a project is successfully completed and its products or services are in demand. This is similar to early acquisition and holding of commodities with a view to trading them later at a higher price. Both ICOs and crowdfunding enable a global reach of investors, as typically investment is marketed over the Web. The difference between an ICO and traditional crowdfunding is that the latter is regulated in most major countries, including the USA, UK and the EU, while ICOs often go unregulated, at least for now. Buying shares or future services via a regulated crowdfunding operator or a private equity sale presents the challenge to realise the returns before the company is sold or goes public. Early-stage investments are often priced based on a promise of future profits rather than existing revenue, which is why agreeing a share price is the first and arguably most difficult task.

Crowdfunding

Crowdfunding is an alternative way of raising finance, usually to launch a start-up business. Traditionally, financing a business, project or venture involved asking a few people for large sums of money. Crowdfunding switches this idea around, using the internet to talk to thousands—if not millions—of potential funders. There are different kinds of crowdfunding:

Reward crowdfunding People invest because they believe in the cause—they might receive a reward, such as an honorary mention, a card or the right to use a service. For example, in the crowdfunding of a smartphone company, people can receive a smartphone at a discounted price.

Debt crowdfunding Investors receive their money back with interest. This activity is also called Peer-to-Peer (p2p) lending; it allows for the lending of money while bypassing traditional banks.

Equity crowdfunding People invest in an opportunity in exchange for equity. Money is exchanged for a shares, or a small stake in the business, project or venture.

In the UK, reward crowdfunding is not regulated. Debt and equity crowdfunding are authorised and regulated by the financial regulator Financial Conduct Authority. In the USA, equity crowdfunding is equally regulated and requires registration with the Security Exchange Commission (SEC).

Because the tokens of an ICO are usually traded on exchanges, buying and selling can be achieved rather quickly; this is why tokens are so attractive for investors. Additionally, investors may at the same time become users of the project financed by the ICO, thus achieving better customer loyalty, and an opportunity is created to finance networks. The high volatility of the tokens is, of course, less attractive.

With ICOs creating an open market, regulators are increasingly concerned about participation by naive consumers and the potential to undermine market stability. The question has moved on from whether ICOs should be regulated, to how should they be regulated? Some countries—China, South Korea and Pakistan for example—have taken draconian measures and banned them outright. Why? Well, research shows that a whopping 78% of ICOs are either fraudulent [45] or fail within three months of funding [106], making the move to ban them more understandable. However, funding new businesses is essential for economic development and raising finance is a major challenge for start-ups or small and medium-sized enterprises. A fully functioning ICO market could be a perfectly legitimate and successful way to overcome this.

What Is the Howey Test?

Following the US Supreme Court's Howey case in 1946, the so called Howey test applies to determine if an *investment contract* exists among parties, regardless of how the contract has been phrased. If the contract is the *investment of money* in a *common enterprise* with a *reasonable expectation of profits* to be derived from the *efforts of others*, then it could be considered, by the US courts, an investment contract—included in the definition of securities. Issuance of securities, and associated activities such as marketing, sale (in primary and secondary markets) and promotion, requires registration with the SEC. To prevent legal repercussion when running an ICO, or trade cryptoassets in the USA, performing the Howey analysis is practical way to initially assess status of the cryptoassets [55].

At this stage, it's important to point out that regulation of ICOs often depends on what the tokens are being used for. Tokens tend to have three functions:

Security tokens Allow participation in a venture or an investment, for example the company shares or a derivative financial product, or a debt.

Utility tokens Represent a pre-sale of goods or services, similar to buying mobile phone future air-time. This is also like reward crowdfunding; BAT token falls into this category (see Chapter 5).

Tokens as payment These play a crucial role in open network security and are fundamentally different from the other two categories of tokens. They are a blockchain-based security, depository receipt or voucher. These include the standard cryptocurrencies like Bitcoin and Ether.

Depending on where they are issued, each type of token is regulated differently—see common classification among various countries in Fig. 9.1. Current financial regulation can be interpreted and applied to security tokens. A security token is a security regardless of how it is issued. In a number of countries, a company must be authorised by or registered with the financial regulator before it can run a security-based ICO. Information about the sale of

Country	Payment or exchange	Investment	Grant access
UK	Exchange token	Security token	Utility token
Malta	Virtual financial asset	Financial instrument	Virtual token
ESMA	Payment cryptoasset	Investment cryptoasset	Utility-type cryptoasset
Switzerland	Payment token / cryptocurrency	Asset token	Utility token
US	Payment token / cryptocurrency	Security token	Utility token

Fig. 9.1 Token/cryptoasset cryptoasset classification frameworks in the jurisdictions discussed in this paper

the tokens needs to be accurate, and regardless of how the money is collected, cryptocurrency or fiat, at a minimum, laws combating money laundering, know your customer and funding of terrorism apply. Registration also applies to companies marketing and promoting the sale.

Rules covering the issue of tokens not directly classified as securities are much less uniform. These tokens are not usually regulated from a financial point of view, although services associated with the token could be. For example, if a token represents a timeshare contract, a legal contract is required to link the token to the use of the property.

Neither the UK nor the USA require utility tokens to be regulated; the USA and retrofits security law to the ICO market (see Chapter 8), while the UK is about to issue new regulation, security tokens regulated by the FCA [41]. Several ICOs have bypassed regulation by claiming to have produced a utility token, when arguably a security token has been issued. The SEC has been proactive in prosecuting these cases; it imposed $250,000 as penalty against Airfox and Paragon Coin Inc for conducting ICOs in 2017 without registration or exemption. Airfox raised $15 million to develop an ecosystem with mobile applications for users in emerging markets to earn tokens and exchange them for data. Paragon raised approximately $12 million to develop a blockchain solution for the cannabis industry and to work towards legalisation of cannabis. Moreover, the companies needed to compensate harmed investors who purchased tokens in the illegal offerings. Finally, both companies had to register their tokens with the SEC [48].

To distinguish between security and utility tokens, it's necessary to understand investors' rights. If an investor can simply gain access to a service, for example an email system, this is considered a utility (it's the same if someone is playing an online game and has to buy tokens to advance). If the tokens entitle them to vote, or to future profits of the business, or interest on the capital, then this is more than likely to be a security token.

Several countries have issued guidance: in the USA, the SEC has issued a guide to companies and investors to determine when tokens are securities [55]. In the EU, the European Securities and Markets Authority (ESMA) has developed a decision tree, that not only helps to determine the function of a token, but also which EU regulation applies (see Fig. 9.2). Malta has defined a questionnaire as a test to determine if registration is needed with *The Malta Financial Services Authority (MFSA)* [49].

Finally, it is worth spending a few words on the status of fiat-backed stable coins, such as Facebook Libra (see Chapter 5), which by means and purposes constitutes a new form of electronic money. These are typically regulated: in the EU, the E-money directive applies together with money laundering and counterterrorism finance regulation. In the USA, the the Department of Justice has investigated Tether (see Chapter 5) for market manipulation.

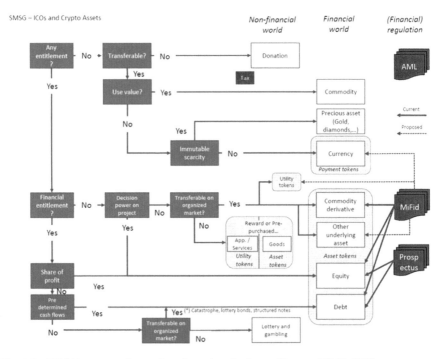

Fig. 9.2 ESMA's test to determine the role of tokens (*Source* ESMA [45])

Still Interested in Tokens?

Assuming you haven't been put off buying tokens, it's important to find out whether the issuing company you are interested in is authorised by the regulator (if it has, it's reasonable to assume the sale information can be trusted). In 2013, in the early days of ICOs, several companies bypassed regulations by requesting donations for their enterprise, a ploy made easier by the Internet. In principle, asking for donations as a source of funding is itself not against regulation. A number of companies rely on donations to survive: Wikipedia is one of the most prominent. However, using it to find a way of avoiding regulation is becoming more difficult, as regulators are becoming more proactive. In the USA, the Securities and Exchange Commission (SEC) has been proactive in prosecuting companies that try to do this.

The Status of Cryptocurrencies

In Chapter 2, we saw that cryptocurrencies don't fit into the economic definition of money and, at the time of writing, no country has declared any cryptocurrency legal tender, and some countries have banned them—see Fig. 2.5. If cryptocurrencies are not money or legal tender, what are they? We have already seen how new cryptocurrencies have been created but this is not sufficient to determine their legal status. The option is limited: cryptocurrencies can either be classified as a form of private money, or commodities, or a tradable asset.

The difficulty is that even within the borders of one country, cryptocurrencies can be viewed in a number of different ways. Naturally, central banks want to ensure they don't get entangled in money laundering scandals; however, governments are keen to levy taxes where legitimate business uses cryptocurrencies. Cryptocurrencies are labelled virtual currencies in the EU and commodities in the USA, and when they are not banned, everything in between. This classification enables the levy of taxes.

In a small number of countries, cryptocurrencies are accepted as a means of payment. In the Swiss Cantons of Zug, cryptocurrencies are even accepted as payment by government agencies, and the Isle of Man, Mexico and Japan also permit the use of cryptocurrencies as a means of payment alongside their national currency. On April 2017, Japan issued a Payment Services Act where cryptocurrencies are defined as virtual currencies.

Taxation

Tax matters if you are going to dip your toe in the world of cryptocurrencies or tokens. Whether gains are made from mining or selling cryptoassets, they are treated in several jurisdictions as selling shares and are taxed accordingly. If, on the other hand, people get paid in cryptocurrency, then income tax applies. Interestingly, the European Court of Justice (ECJ) declared that gains from cryptocurrency investments are not subject to value-added tax in the EU member states. Several countries seem to be in the process of devising taxation rules. Mining of cryptocurrencies is normally also exempt from taxation; in Russia, however, mining that exceeds a certain energy-consumption threshold is already taxable.

Exchanges and Secondary Markets

For an ICO, when after the initial distribution tokens are traded in secondary markets, what is the regulation that applies? Once again, in several countries, trading of securities can only take place on regulated exchanges. As most are online, they require a licence to operate from the country where they trade, and need to comply with the law in all the countries where services are sold. The company behind an exchange, which receives money to keep the tokens on behalf of investors, needs to comply with money laundering, know your customers and anti-terrorism legislation. They also need to follow best practice that enables fair trading and regulation of the ancillary services associated with distribution of information about the tokens. Consumers have more protection when buying from licensed exchanges.

In the EU, exchanges need to comply with the Fifth Anti-Money Laundering Directive (AMLD5), which will need to be transposed into member states' own legislation by January 10, 2020. This new directive extends previous regulation and brings virtual currency exchange platforms and custodian wallet providers within the scope of the EU's anti-money laundering requirements.

There is more room for regulatory improvement for the custody of private keys, where a platform can offer this kind of service. The impact of loss or theft of keys can be disastrous to users when no appropriate steps have been taken for the recovery. Research shows that 62% of large platforms retain control over customer funds, as opposed to 30% of small firms. Crucially, two-thirds of specialised custodial exchanges do not have a refund procedure in place for when customer funds get lost or are stolen [99].

Cryptocurrency exchanges allow customers to trade cryptocurrencies or tokens for other assets including conventional fiat money. There are also market makers that take bid-ask spreads as transaction commissions for their services, or charge fees as a matching platform. In Japan, the sale, purchase and exchange of virtual currencies are regulated and require registration with the Financial Services Agency of Japan (JFSA). In January 2018, Coincheck, a Japanese exchange, suffered a loss of $500 million by the hands of hackers. The JFSA acted swiftly by issuing punishment notices and halting business.

Cryptocurrency exchanges are integral to the cryptoasset ecosystem. According to the Financial Stability Board (FSB), an international body that monitors and makes recommendations about the global financial system, cryptocurrency exchanges have not, as yet, raised any major concerns. However, their impact on consumer protection and money laundering has prompted regulatory intervention.

If you decide to buy currencies from an exchange, make sure the exchange is properly licensed; if it isn't, you can wave goodbye to any form of legal redress. Deploying decentralised exchanges (see Chapter 5) will provide more anonymity and responsibility for the custody of the assets; there is, however, no legal redress in case private keys are lost or stolen. Decentralised exchanges or DEX present the biggest headache for regulators. As it can be difficult to determine who is in charge, and often participants have no obligation to disclose their identity, money laundering and anti-terrorism regulations are difficult to enforce.

Technology

Finally, if we view blockchain as a back-end technology (see Chapter 6) that enables a tamper-proof record of information collectively curated by a community of participants, do we really need to regulate it at all?

Blockchain has become a buzz word, used as a PR and marketing tool by some companies. The challenge for the end-user will be familiar to consumers of organic food: How do I distinguish the organic carrot from one covered in pesticides? A guarantee via a label would be the obvious answer. So, do we need something similar for blockchain technology? Do we need legislation that protects the buyer, so that when someone is searching for a blockchain solution because they really believe in the technology and its benefits, they will get what they are looking for? If the answer is 'yes', we need that, then somebody needs to regulate blockchain technology.

Malta is one of the few countries that have taken the view that regulation is necessary, creating the *Digital Innovation Authority* (*MDIA*), the primary authority responsible for promoting all governmental policies [35]. The *Innovative Technology Arrangements and Services Act* (*ITAS*) enables the creation of a register of service providers, including Systems Auditors and Technical Administrators with the task of certifying blockchain technology, smart contracts and related applications, including Decentralised Autonomous Organisation.

Summary

Cryptocurrencies have unleashed a parallel and unfamiliar financial ecosystem. Regulators need to take a decision on how much existing regulation can be applied, and where new regulation is needed. One size does not fit all: some countries have taken a positive approach, with legislation aimed at attracting new businesses; others have chosen to retrofit current regulation. People investing and deploying crypto-systems have to be aware of these regulations; failing to do so could have a very negative impact on their business.

Regulation will continue to be a feature as blockchain and cryptocurrencies continue to develop. It is likely that more regulation and education are needed to protect consumers as the market grows.

10

The Future of the Blockchain

History is littered with predictions that have proved to be spectacularly wrong. In 1876, William Orton, President of Western Union said 'This "telephone" has too many shortcomings to be seriously considered as a means of communication', contemptuously dismissing Alexander Graham Bell's new invention.

Today, there are many sceptics who are equally ready to dismiss cryptocurrencies and blockchain technology as little more than a passing fad. However, Christine Lagarde, the highly respected managing director of the International Monetary Fund, is not one of them. Recalling William Orton's words [91], in 2108 she said blockchain technology could be a modern-day equivalent: 'I think the role of the disruptors and anything that is using distributed ledger technology, whether you call it crypto, assets, currencies or whatever … that is clearly shaking the system' [50]. Lagarde's message was clear: the full impact of blockchain may take time to see, but see it we undoubtedly will.

What Does the Future Hold for Blockchain?

You can be interested in blockchain technology because of the underlying technology, or because of its commercial possibilities, or because of its opportunity changes how businesses operate, or for ideological reasons. The original technology, born with Bitcoin, is based on sound technical principles, and its evolution has lead to a rich and varied digital ecosystem. At the time of writing, there are more than two thousand cryptocurrencies or digital currencies in

© The Author(s) 2020
M. G. Vigliotti and H. Jones, *The Executive Guide to Blockchain*,
https://doi.org/10.1007/978-3-030-21107-3_10

existence. Not all of them follow the bitcoin model; some are created through mining, others by a single act of issuance, with a third hybrid model the preferred method for another group. Some only allow the transfer of coins, while others, like Ethereum or Tron, are more versatile, providing a platform for smart contracts. What underpins all of these different cryptocurrencies, however, is decentralisation, or not having third party or government involvement as overseer or controller.

In this book, we have explored the evolution of blockchain via permissioned blockchain for enterprises that like to have more control over who has access to the ledger. Examples of these kinds of blockchain were discussed: supply chain, trade finance and land registry. We also had the courage to discuss when blockchain should be avoided, and why: there are challenges with personal data and voting systems.

Having already examined in detail the development of blockchain and its growing impact on business, we must now consider what the future holds for this technology. What will allow it to flourish? What might inhibit its growth? As evident as it might sound, encouraging a greater uptake of blockchain, whether as a payment system or as an opportunity to support early ventures, is crucial to its future. The appeal of cryptocurrencies is growing rapidly, particularly among younger, tech-savvy types. According to a survey in 2017 by the venture capital firm Blockchain Capital, 30% of 18 to 34-year-olds in the USA would rather invest $1000 in Bitcoin than in government bonds or stocks. The same study also showed that 42% of millennials had heard about bitcoin, compared with 15% awareness among those aged 65 or older [57]. [Bloomberg news article, 8.11.2017, nearly a third of millennials says they would rather own bitcoin than stocks.]

Advancements in technology could have both a positive and a negative impact on blockchain. One of the selling points of the technology is that cryptographically secured distributed ledgers are, under normal circumstances, virtually unbreakable. However, quantum computing, with its potential to disrupt public key cryptography, could compromise the security of blockchain-based systems and leave them vulnerable to threats not envisaged when they were designed. We will look at this in more detail later in this chapter.

While this could be seen as a dark cloud spoiling a rather distant horizon, it is unlikely to halt the progress of blockchain. The financial and charitable sectors are two prominent and enthusiastic adopters of the technology, as are supply chain companies and the creative industries. The outlook is exciting, although some bumps in the road can't be ruled out.

Have Cryptocurrencies Peaked?

With more than two thousand cryptocurrencies in circulation, the crypmarket looks overcrowded. About 40% are dead currencies, with a value of less than a penny. There are a number of reasons for this, including: a hard-hitting bear market in the second half of 2018; scam ICOs resulting in heightened investor caution; and tougher regulation that has killed weaker projects.

Over the next few years, it's likely there will be a fall in the number of cryptocurrencies and ICOs, leaving the field open to bigger players, particularly those with a product that meets the needs of the market. Facebook, for example, has signalled its intention to launch a cryptocurrency in 2020. Libra, a so-called stable coin, whose value would be pegged to a basket of major fiat currencies, will be built into Facebook, Messenger and WhatsApp, allowing users to send and receive money via messages. Facebook claims the currency will provide financial services to an estimated 1.7 billion people worldwide who don't have bank accounts. Facebook is building a consortium, based in Geneva, to manage the hard-currency reserves of Libra. Currently, 28 prospective founding members are enlisted, though a more optimist prospective suggests the consortium would reach 100: including financial firms (Visa, Stripe), online services (Spotify, Uber), cryptocurrency wallets (Anchorage, Coinbase), venture capitalists (Andreessen Horowitz, Union Square Ventures) and charities (Kiva, Mercy Corps).

Facebook won't have failed to notice how the banking sector has, up to now, managed to remain largely untouched by the disruptor model that has shaken up so many sectors. Digital payments, such as Apple Pay or PayPal, are, in effect, merely piggybacking the three major credit card companies, which between them take a quarter of the $30 billion annual market [54].

Mark Zuckerberg's all-conquering company is hungrily eyeing the opportunity to tap into its 2.4 billion users, many of whom will be using traditional banking methods, and convert them to Libra. It's a mouth-watering prospect for the Silicon Valley behemoth, but there are many regulatory hurdles to clear before it becomes the next major player in the cryptocurrency market.

As for the existing market, two players continue to stand out: Bitcoin and Ethereum. It's not possible to predict what will happen when both are fully mined as they one day will be, but it seems unlikely either will lose their popularity; indeed, in the short term, Bitcoin will almost certainly maintain its dominance, based on its current popularity and its status as the most trusted cryptocurrency in the market.

Despite this popularity—rather, because of it—Bitcoin faces issues with the number of transactions it can process at any given time (currently an upper limit of seven per second), although this is being addressed by the Bitcoin community. Indeed, a project known as the Lightning Network has been developed on top of the Bitcoin blockchain [15], that has the potential to enable millions of transactions per second. Although not integral to Bitcoin, Lightning Network will be part of its core blockchain. With the agreement of participants, transactions will be conducted 'off-chain' with the ledger updated later (although transactions will continue to appear instantaneous to participants). Currently, still under development, Lightning Network has, nevertheless, seen a steady increase in daily transactions.

Ethereum, although not as well-known as bitcoin, continues to lead the way in smart contract or distributed application platforms, despite the fact that 60% of contracts have been shown to have serious vulnerabilities (see Chapters 4 and 8). Ethereum is also moving from the consensus algorithm PoW to the environmentally friendly PoS, which, if successful, will lead to an increase in the number of transactions processed.

Looking ahead, it's likely that regulators will prevent tech giants like Facebook from becoming too powerful and overwhelming its competitors through a blockchain-based payment system and seek to beef up their efforts to shape ICO markets. As we have already seen in Chapter 9, both Malta and Switzerland have used regulation to attract investment from companies who might otherwise have been wary of operating in a 'wild-west' environment.

As we consider the future of cryptocurrencies, it is worth keeping in mind that the total capitalisation of this developing market is still relatively small. At its peak in December 2017, it was close to $1 trillion, about the same size as Amazon. At the moment, activity from cryptocurrencies and cryptoassets is not considered a risk for the global financial stability [39]. If the market continues to grow, increased regulation is very likely with countries working together to provide a uniform and harmonised set of rules. Among other things, regulators will be looking for clarity about the status of tokens; proof that exchanges operate fairly; and evidence of a structured approach to providing legal status to smart contracts. Deployment of specialised task forces to prevent money laundering and funding of illegal activities, and terrorism will also be high up the agenda.

Quantum of Solace, or a Major Threat to Blockchain?

As we saw earlier in this chapter, *quantum computing*, which is based on *quantum mechanics*, has the very real potential to threaten the security of blockchain technology. While it is nearly impossible to launch a successful attack on a blockchain using a standard computer, that isn't, at least theoretically, the case with a quantum computer. It is then much more likely that a hacker using a sufficiently powerful quantum computer could identify a private key. Once achieved, they could then read messages, change the content of encrypted information, or impersonate a key holder and sign Bitcoin (or other cryptocurrencies) transactions to steal bitcoins from other accounts. However, the potential for it to actually happen depends on how cryptography is deployed on a blockchain. Taking Bitcoin as an example, it would be possible for an attacker who gained possession of a private key used for signing transactions; this would mean someone else could take control of your bitcoins. Thankfully, it wouldn't be possible for them to double spend coins or change the ledger because a quantum computer wouldn't help break the cryptographic hashes (see Chapter 4), but only enable reading them.

Although quantum computers aren't commercially available, research into quantum computing and computers is very active and attracts attention from many different directions. A number of companies, including Google, IBM, Microsoft, Intel, Atos, Baidu and Alibaba, are busy working on their development, while several countries all around the world, including Australia, Canada, China, EU, Japan, Netherlands, Russia, Singapore, UK and the USA, have launched national quantum technologies programs [107]. It is believed that a working quantum computer is not far away [107], yet it is not clear if the benefits will be as great predicted.[1] It is difficult to predict when a commercial and affordable model, and even more so when one powerful enough to attack RSA [101], will be available.

Meanwhile, the best and brightest minds are working hard to find a solution to the theoretical threat posed by quantum computers through the development of *post-quantum algorithms* or *quantum-resistant algorithms*. Some blockchains even claim to have these ready for deployment, although this should be viewed with a large degree of scepticism as deploying quantum

---

[1]Quantum computers have the potentiality to be very fast, but will be of low precision, whereas the classical computers are slow, but have a far higher precision. To achieve the same precision with quantum computers, quantum calculations would need to be repeated multiple times, after which the results should be averaged; this would annihilate the advantage of speed.

resistant algorithms requires a highly specialised workforce currently mostly confined to national research laboratories and academia.

But a much more important point of concern is: if—or more likely when—powerful quantum computers become widely available, they will more than anything pose a direct and serious threat to the standard banking system, with its reliance on Public Key Infrastructure; any concerns about blockchain security in comparison will seem almost irrelevant.

Will Central Banks Get on Board the Cryptocurrency Train?

With the use of cryptocurrencies becoming more widespread, might we soon see banks issuing their own digital money? The use of cash is steadily declining [67], so it would make sense for central banks, some of whom are already trialling blockchain technology, to issue a privacy-preserving cryptocurrency. Before exploring this possibility in more detail, it is worth thinking about the role of central banks beyond its remit to issue money. Many central banks have an independent role in determining monetary policy as well as oversight of commercial banks. Their main aim is to keep a nation's fiat currency stable and control inflation.

Central banks issue and control a 'two-tier' source of money: physical money for use by the general public and digital money created by commercial banks. If a central bank was to issue a digital currency, with or without blockchain technology, it would function much like cash: after issue it would circulate between banks, non-financial firms and consumers without the need for a central bank to keep track of the currency or adjust balances.

A key challenge facing any central bank wanting to issue a digital currency would be the need to replace this two-tier system. It's unlikely that Bitcoin or any other permissionless peer-to-peer currency will displace national currencies because central banks:

- Need to control the issue of a currency to maintain stability.
- Would find it difficult to eliminate the existing 'two-tier' source of money.

The Role of a Central Bank

Central banks manage economic growth by:

- Controlling the money supply.
- Buying and selling securities (this changes the amount of cash on hand without changing the reserve requirement).
- Setting interest rates they charge their member banks (commercial banks) (generally, raising rates slows growth and prevents inflation; lowering rates stimulates growth and helps prevent or shorten a recession).

Let's consider alternative ways of a central bank issuing a currency on the blockchain. One way is to issue a currency whose value is pegged to the national fiat money, so would essentially be a digital version of cash. The blockchain would also be a means validated and settled, and would hopefully improve interbank transfers, effectively offering the central bank the power of supervision—but eliminating the two-tier money system. This could deploy a permissioned blockchain (see Chapter 5) whereby participants are limited and must be granted access to participate in the network.

The system of two-tier money helps central banks maintain a supervisory role over commercial banks. If central banks start issuing cryptocurrencies that are available to the public and commercial banks, it would lead to the man-in-the-street holding an account with a central bank, also known as *central bank deposited currency accounts*. Because deposits in central banks are safer, a possible outcome is the increase of central banks deposits by individuals and a substantial reduction of commercial bank deposits. A realistic financial risk is then the weakening of commercial and retails banks, who play an important role in funding businesses and growing the economy. An unavoidable consequence would be that central banks start competing with retail or commercial banks. Why this is not a desirable outcome? Mostly as a matter of policy, as competition with commercial banks could undermine the supervisory role. Secondary, most central banks are not equipped to manage bank accounts. A considerable amount of resources would be necessary to enable the transition.

While there is an active debate about the future role of central banks and whether they should offer deposited accounts, in the short term it's unlikely there will be a significant change in their role. Research shows that only a few central banks are likely to issue national digital currencies in the next decade, although it's unlikely they will be based on blockchain technology [64]. Several central banks are researching, and a few are also experimenting: the Swedish

Central Bank (Sveriges Riksbank) is working on issuing the *e-krona*, the digital version of the krona, as an alternative form of money as cash usage is declining [52]. In 2017, the Central Bank of Uruguay began a pilot programme to issue the *e-Peso*, to improve financial inclusion [64]; both these examples do not seem to deploy blockchain technology. The Bank of France and the National Bank of Cambodia deployed the blockchain technology to improve payments systems rather than the issuance of cryptocurrencies [52].

As long as the role of the central bank is to maintain financial stability, central banks will inevitably want to maintain a tight control over the money supply, and this would mean not fixing the number of coins to be issued, and possibly not deploying blockchain technology.

The Future of Cross-Border Payments

In Chapter 5, we looked at the difference between currencies used to facilitate payments, such as Ripple and Stellar, and other cryptocurrencies. The former are used in the Western world to speed-up cross-border payments, reduce costs and create a bridge for places with a poorly connected banking infrastructure. However, they don't easily address the problem of making cross-border payments in countries with very poor or no banking infrastructure. Cryptocurrencies like Bitcoin, which deploy an alternative payment channel, could in principle be a solution to improve cross-border payment systems. The challenges for mass adoption, which requires the development of an ecosystem of operators willing to invest, are regulation, scalability and competition from alternative mobile payment systems, such as M-Pesa, which today serve the unbanked.

M-Pesa and other pure mobile payment systems abide to regulation by obtaining a license to transfer money and process payments, without being a bank. They have developed a financial ecosystem of micro-lending, credit and savings. The penetration of these mobile operators in Africa and Asia is impressive. As a result, it will be challenging, but not impossible, for operators deploying cryptocurrencies to get a substantial share of the market.

Private coins could become more widespread in countries where political instability exists and where individual freedoms are limited. However, unless a parallel economy based on a specific cryptocurrency takes hold, individuals will rely on crypto-exchanges to get the national currency. If exchanges are declared illegal by a hostile government, cryptocoins would become worthless.

The Dream of a Decentralised Society

The early blockchain rhetoric was centred around technical decentralisation, disintermediation of the state, and redistribution of power from well established legal, social and financial institutions [95]. Ironically, ensuring secure direct commerce with far-away parties, and fair redistribution of power to the masses, it is now very close to the original democratic goals that led to modern Western democracies.

Rather than considering decentralisation as disintermediation, is there a way in which technology can actually complement or even improve processes to provide some better form of democracy? Some institutions, such as the courts, are very difficult to replace when they work well—as discussed in Chapter 8.

Considering the land registry discussed in Chapter 6, clearly identified parties, not anonymous ones, can deploy a blockchain to transfer properties without solicitors, yet a traditional court system can settle any non-trivial dispute. Contrary to shared belief in some blockchain circles, technology alone does not deliver disintermediation, but compelling economic issues can go further in delivering innovative solutions outside traditional institutions, as the success of mobile payments has clearly demonstrated.

A

Useful Resources

Bitcoin

Bitcoin original project Complete information about the Bitcoin project can be found at the domain anonymously bought in August 2008 https://bitcoin.org/. The website is very well curated and kept up to date by the community.

Transaction Explorer Several companies make the Bitcoin ledger publicly available, see, for example, https://blockchair.com/bitcoin or https://live.blockcypher.com/btc/. None of these are endorsed by the Bitcoin community.

Lightning Network The Lightning Network is an independent technological development with the goal to deliver high-volume micro-payments on the Bitcoin network. For more information, see https://lightning.network/ (see discussion in Chapter 10).

Bitcoin Foundation The Bitcoin Foundation is an independent organisation with the mission to promote Bitcoin, see https://bitcoinfoundation.org.

Cryptocurrencies

Below we provide details (a brief history of platform, key features, where to find the open-source code, and how to track transactions in real time) of the cryptocurrencies discussed in Chapter 5.

© The Editor(s) (if applicable) and The Author(s), under exclusive licence **173**
to Springer Nature Switzerland AG 2020
M. G. Vigliotti and H. Jones, *The Executive Guide to Blockchain*,
https://doi.org/10.1007/978-3-030-21107-3

Litecoin Litecoin (LTC) (https://litecoin.org) was launched in October 2011. The first so-called altcoin, it was developed by Charlie Lee using a modified version of the Bitcoin blockchain. Differences from Bitcoin include:

a. Cutting average time for miners to generate a new valid block to 2.5 minutes by using an alternative algorithm for guessing the nonce in the Proof of Work
b. Setting total number of coins generated at 84 million
c. Halving the block reward every 840,000 blocks
d. Creating an initial block reward of 50 Litecoins.

Litecoin is one of the most capitalised cryptocurrencies and there are more than 50 million coins in circulation.

Location Code Litecoin is open source; software can be found at https://github.com/litecoin-project/litecoin.

Transaction Explorer Litecoin blockchain transactions can be monitored in real time at https://chainz.cryptoid.info/ltc.

Bitcoin Cash Bitcoin Cash (BCH) (https://www.bitcoincash. org), originated from Bitcoin; software was changed to create another blockchain (see Chapter 4), with the aim of doubling the number of transactions processed.

It would facilitate Bitcoin to be used as payment system. Bitcoin and BitcoinCash shared the same transactions until block 478558 was mined. From block 478559 the two separated and now trade at different prices. BitcoinCash's decision to double the block size led to the split in August 2017. Today, BitcoinCash is among the most capitalised cryptocurrencies.

Location Code BitcoinCash is open source, and the code for the software is located at https://github.com/bitcoincashorg/bitcoincash.org.

Transaction Explorer BitcoinCash blockchain transactions can be monitored in real time at at https://github.com/bitcoincashorg/bitcoincash.org.

Zcash Launched in 2016 by Wilcox-O'Hearn, Zcash (ZEC), (https://z.cash) shares Bitcoin's codebase; it preserves user anonymity and privacy. Like Bitcoin, the ledger can be pub- lic, but only with the permission of sender and receiver.

Without permission, the amount of coins sent between users and their addresses will not show on the ledger. Z-cash uses a technique called 'Zero knowledge proof' that allows users to share selected details, for compliance or audit, without having to reveal full details. The total number of coins in circulation has been set at 21 million, and the block-reward started at 50 ZEC, and halves very 840,000 blocks.

Location Code Zcash is an open-source cryptocurrency; software can be found at https://github.com/zcash/zcash/projects.

Transaction Explorer Z-cash blockchain transactions can be monitored in real time at https://zcash.blockexplorer.com.

Monero Monero   (XMR)   (https://www.getmonero.org): launched in 2014, Monero prevents third parties from seeing address and amount of Moneros exchanged on the ledger.

Privacy is achieved through use of a cryptographic technique called Ring Signatures. Users can look at the Monero ledger to check the amount of coins transferred, the time when the block was created, and the fees paid to the miners. However, identifying addresses is almost impossible. The total supply of Monero, 22,482,672.60, is expected to be reached in 2050; blocks are created on average every two minutes.

Location Code Monero is an open-source cryptocurrency and the software can be found at https://github.com/monero-project/monero.

Transaction Explorer Monero blockchain transactions can be monitored in real time at https://moneroblocks.info.

Dash (DASH) (https://www.dash.org) was launched in January 2014 as 'Xcoin' by Evan Duffield. Rebranded as Darkcoin, it became Dash in March 2015. Duffield modified the Bitcoin code to perform fast, untraceable transactions.

The maximum supply of coins is set to 22 million. Coins are mined using a modified version of the Bitcoin Proof of Work. Dash operates a decentralised autonomous organisation, meaning the governance structure relies on a set of computers that act as shareholders, voting on proposals for improving Dash's ecosystem. Anybody can run a masternode, but ownership of 1000 Dash is required. At the time of writing[1] there are 4800 masternodes. As with Bitcoin, the blockchain runs with standard nodes and miners. It is very popular in Venezuela (see https://discoverdash.com/stats).

Location Code Dash is an open-source cryptocurrency and the software can be found at https://github.com/dashpay/dash.

Transaction Explorer Dash blockchain can be monitored in real time at https://insight.dash.org/insight.

---

[1] Also see https://www.dashninja.pl (accessed April 26, 2019).

Ethereum Ethereum (ETH) (https://www.ethereum.org) is the second-most popular blockchain and Ether is the second-most capitalised cryptocurrency. The idea of Vitali Buterin, who set out his vision in 'Ethereum: The Ultimate Smart Contract and Decentralised Application Platform'.

https://Ethereum.org was registered on 27 November 2013; the following July crowdfunding in bitcoin began lasting 42 days and promising each subscriber between 1337 and 2000 ETH per bitcoin. Crowdfunding raised $18.4 million. Ether was initially created and distributed to crowdfunding subscribers; similar to Bitcoin, most Ether is still to be mined. Maximum supply has been set at 100 million. Launched in 2015, the platform quickly became popular because of some functional advantages over Bitcoin. For example, developers can write code that is executed on the Ethereum chain. Known as 'smart contracts', these snippets of code, which result in software applications that deploy the Ethereum platform, are often referred to as *Distributed Applications* (DApps).

Ethereum is also popular because it has lowered the barrier or entry in creating new crypto currencies. By writing at most a few hundred lines of code via smart contracts a new cryptocurrency can be created in a day or so. Although different from Ether, these new crypto-coins are still mined by Ethereum miners.

Location Code Ethereum is open source cryptocurrency and the software can be found at https://github.com/ethereum.

Transaction Explorer Ethereum blockchain transactions can be monitored in real time at https://etherscan.io.

Ethereum Classic Ethereum Classic (ETC) (https://ethereum classic.org) was created in July 2016 from Ethereum at block 1920000. Since then, the two crypto currencies have developed separately.

To understand how Ethereum Classic came about See Box 'The Darkest Day in Ethereum's History'. Ethereum Classic is now a completely different platform. It has the facility to deploy smart contracts. The community, however, is much smaller, and as a consequence the platform development is much slower.

Location Code Ethereum Classic is an open-source cryptocurrency and the software can be found at: https://github.com/ethereumproject.

Transaction Explorer Ethereum Classic blockchain transactions can be monitored in real time https://gastracker.io.

Cardano Cardano, created in 2015, is a blockchain supporting the Ada coin (Ada) (https://www.cardano.org/en/home). The Cardano platform was set up to run financial and decentralised applications. Cardano is being developed in conjunction with a team of leading academics and engineers to ensure the technology is secure and has the same scientific rigour that is applied to mission-critical systems.

> **Location Code** Cardano is an open-source cryptocurrency and the software can be found at: https://github.com/input-output-hk/cardano-sl.
>
> **Transaction Explorer** Cardano blockchain can be monitored in real time https://cardanoexplorer.com.

Tron Created in 2018, Tron (TRX) (https://tron.network) uses Ethereum software and Proof of Stake consensus algorithm. It will enable developers to create smart contracts and decentralised applications.

> **Location Code** Tron is an open-source cryptocurrency and the software can be found at https://github.com/tronprotocol.
>
> **Transaction Explorer** Tron blockchain can be monitored in real time https://tronscan.org/#.

Ripple Ripple (XRP) (https://ripple.com) was created by Jed Mc-Caleb in May 2011. All participants have a pre-existing re- lationship of trust that doesn't exist on the Bitcoin network. Ripple mostly connects financial institutions to enable cross-border payments where XRP coin is used as a unit of account to determine the conversion rate. Recently Ripple has developed more user-friendly products to increase adoption.

> **Location Code** Ripple is an open-source cryptocurrency and the software can be found at https://github.com/ripple.
>
> **Transaction Explorer** Ripple blockchain transactions can be monitored in real time https://xrpscan.com.

Stellar The Stellar platform (Lumen) (https://www.stellar.org) was funded in 2014. It aims at facilitating cross-border payments while maintaining consistent low fees. Stellar can handle exchanges between fiat-based currencies and between cryptocurrencies. The Stellar foundation that governs the Stellar platform is non-profit. Stellar is also a payment technology that aims to connect financial institutions and drastically reduce the cost and time required for cross-border transfers.

> **Location Code**   Stellar platform is an open-source cryptocurrency and the software can be found at https://github.com/stellar.
>
> **Transaction Explorer** Lumen coins can be monitored in real time https://stellarchain.io.

Basic Attentions Token BAT (BAT) (https://basicattentionto ken.org) is part of the Browser Brave. It is a token created on Ethereum using an ERC20 standard. The project seeks to address fraud and opaqueness in digital advertising. The token aims to correctly price user attention within the platform. Advertisers pay BAT to website publishers for the attention of users. The BAT ecosystem includes Brave, an open-source, privacy-cantered browser designed to block trackers and malware. It leverages blockchain technology to anonymously and track user attention securely and rewards publishers accordingly.

> **Location Code**   BAT is an open-source token and the software can be found at https://etherscan.io/address/0x0d8775f648430679a709e98d 2b0cb6250d2887ef#code.
>
> **Transaction Explorer** BAT coins can be monitored in real time https://etherscan.io/token/Bat.

Tether Tether (USDT) (https://tether.to) Tether was launched in 2014, initially with the name real Realcoin. The value of Tether is pegged to the U.S.\$, meaning that Tether similarly to central bank tries to keeps the price of one Tether very close to \$1. In 2018, Tether moved platform, and currently is resides on the Tron platform, having been previously launched on Ethereum.

> **Location Code**   Tether doesn't need its own blockchain platform: Tron tokens can be exchanged on other platforms. Born originally smart contract for the token's creation can be found in: https://etherscan.io/address/0xd697A61D5FB4e076125e0bE647f902 b02bb3A0F1#code   the Ethereum blockchain.

Transaction Explorer   Because Tron and Ethereum share a few features, the Tether token has now migrated on the Tron platform and the transactions can be monitored in real time at: https://tronscan.org/#/contract/TR7NHqjeKQxGTCi8q8ZY4pL8otSzgjLj6t/transfers.

Enterprise Blockchain

Some prominent consortia in the enterprise blockchain space.

B3i   Born in 2018, B3i (see https://b3i.tech/home.html) is rapidly expanding. The consortium is formed by 40 companies, and the vision is to deliver insurance better solutions for end consumers. The first product, based on blockchain technology, will be available in 2020.

Bankchain   BankChain (http://www.bankchaintech.com/index.php) was formed in February 2017. It is a community of banks for exploring the opportunity to deploy blockchain solutions, with currently 8 live projects and 37 members.

Marco Polo Network   Born in September 2017, Marco Polo Network (https://www.marcopolo.finance/) is a large network to enable trade finance in the world. It is formed by both corporate companies and financial institutions with the goal to create one, shared and connected network that brings all the parties together to the limitation of current practices based on inefficient, paper-based systems.

Enterprise Ethereum Alliance   The Enterprise Ethereum Alliance (EEA) (https://entethalliance.org/) is a membership organisation separate from the Ethereum project. Members can be both organisations and individuals. The EEA operates in the enterprise space and its goal is to drive the use of Ethereum blockchain technology within the enterprise space by delivering technological solutions, training and standards.

Glossary

Agreement an understanding or arrangement reached between two or more parties

bitcoin indicates the actual currency

Blockchain is a system of electronic records that it enables a network of independent participants to establish a consensus around the authoritative ordering of cryptographically 'signed' transactions. These records are made persistent by replicating the data across multiple nodes and tamper-evident by linking them by cryptographic hashes. The shared result of the reconciliation/consensus process—the 'edger'—serves as the authoritative version for these records.

Cryptocurrency virtual currency that deploys a blockchain to settle transactions.

Cryptographic Hash unique identifier of a set of data that is almost impossible to forge.

Distributed Ledger Technology see blockchain

Double Spending transfer a cryptocoin to two different recipients at the same time.

Fiat Money money issued by a central bank, such as the US dollar or the British pound. The word 'fiat' comes from the Latin and means 'Let there be'.

Initial Coin Offering a digital way of raising funds from the public using mostly cryptocurrency.

M. G. Vigliotti and H. Jones, *The Executive Guide to Blockchain*, https://doi.org/10.1007/978-3-030-21107-3

Miner A computer or group of computers that add new transactions to blocks and verify blocks created by other miners. Miners collect transaction fees and are rewarded with new cryptocurrencies for their services.

Smart Contract software application that runs on the blockchain.

Token a digital asset generated on a blockchain that does not reside on its own blockchain and does not require the process of mining to be created.

Virtual Currency A virtual currency is a digital representation of value that is issued neither by a central bank nor by a public authority, rarely attached to a fiat currency, but is accepted by a growing number of natural or legal persons as a means of payment and can be transferred, stored or traded electronically.

Bibliography

1. Acting manhattan U.S. attorney announces forfeiture of $48 million from sale of silk road bitcoins. https://www.justice.gov/usao-sdny/pr/acting-manhattan-us-attorney-announces-forfeiture-48-million-sale-silk-road-bitcoins. Accessed May 5, 2019.
2. An introduction to the uk's interbank payment schemes. http://www.accesstopaymentsystems.co.uk/. Accessed March 20, 2019.
3. Bank of Thailand digital currency scheduled for 2019. https://www.bangkokpost.com/business/finance/1526094/bank-of-thailand-digital-currency-scheduled-for-2019. Accessed May 7, 2019.
4. BBVA issues the first blockchain-supported structured green bond for MAPFRE. https://www.bbva.com/en/bbva-issues-the-first-blockchain-supported-structured-green-bond-for-mapfre/. Accessed May 5, 2019.
5. BBVA leads the way in the use of blockchain, according to Forbes. https://www.bbva.com/en/bbva-leads-the-way-in-the-use-of-blockchain-according-to-forbes/. Accessed May 12, 2019.
6. Constitution of the United States. https://www.senate.gov/civics/constitution_item/constitution.htm. Accessed April 11, 2019.
7. Cryptocurrency investment fund industry graphs and charts. https://cryptofundresearch.com/cryptocurrency-funds-overview-infographic/. Accessed May 6, 2019.
8. The digit economist. https://digiconomist.net/. Accessed April 14, 2019.
9. Emails. https://satoshi.nakamotoinstitute.org/emails/. Accessed April 11, 2019.

10. Global trade finance market size, share, statistics trends with detailed analysis of financing skill, industry applications, overview and forecast 2024. https://www.reuters.com/brandfeatures/venture-capital/article?id=90162. Accessed July 14, 2019.

11. Hsbc settles $250bn of FX transactions using distributed ledger technology. https://www.hsbc.com/media/media-releases/2019/fx-everywhere. Accessed May 12, 2019.

12. Iso/tc 307 blockchain and distributed ledger technologies. https://www.iso.org/committee/6266604.html. Accessed June 30, 2019.

13. J.P. Morgan deploys blockchain with new correspondent banking network. https://www.jpmorgan.com/country/GB/en/detail/1320562088910. Accessed July 14, 2019.

14. Largest Number of Banks to Join Live Application of Blockchain Technology. https://www.jpmorgan.com/global/treasury-services/IIN. Accessed April 29, 2019.

15. Lightning network. https://lightning.network/. Accessed June 17, 2019.

16. M-pesa. https://www.safaricom.co.ke/personal/m-pesa. Accessed April 14, 2019.

17. Mas and SGX successfully leverage blockchain technology for settlement of tokenised assets. http://www.mas.gov.sg/News-and-Publications/Media-Releases/2018/MAS-and-SGX-successfully-leverage-blockchain-technology-for-settlement-of-tokenised-assets.aspx. Accessed May 5, 2019.

18. The official libra white paper. https://libra.org/en-US/white-paper/. Accessed July 10, 2019.

19. Petro. https://www.petro.gob.ve/index_eng.html. Accessed May 7, 2019.

20. Regulatory sandbox. https://www.fca.org.uk/firms/regulatory-sandbox. Accessed May 5, 2019.

21. Santander launches the first blockchain-based international money transfer service across four countries. https://www.santander.com/csgs/Satellite/CFWCSancomQP01/en_GB/Corporate/Press-room/Santander-News/2018/04/12/Santander-launches-the-first-blockchain-based-international-money-transfer-service-across-four-countries-.html. Accessed May 12, 2019.

22. Venezuela to lop five zeros off its currency. https://www.ft.com/content/3edcdf1a-90fc-11e8-b639-7680cedcc421. Accessed May 7, 2019.

23. Why blockchain's smart contracts aren't ready for the business world. https://www.gartner.com/smarterwithgartner/why-blockchains-smart-contracts-arent-ready-for-the-business-world/. Accessed June 15, 2019.

24. Computer misuse act 1990. https://www.legislation.gov.uk/ukpga/1990/18/contents, 1990. Accessed May 10, 2019.

25. B-money. http://www.weidai.com/bmoney.txt, 1998. Accessed May 20, 2019.

26. Community framework for electronic signatures. https://eur-lex.europa.eu/legal-content/EN/TXT/?uri=CELEX:31999L0093, 1999. Accessed March 20, 2019.

27. Directive on electronic commerce. https://eur-lex.europa.eu/LexUriServ/LexUriServ.do?uri=CELEX:32000L0031:en:HTML, 2000. Accessed March 20, 2019.

28. Bitcoin—open source p2p money. https://bitcoin.org/, 2008.

29. The trust machine. https://www.economist.com/leaders/2015/10/31/the-trust-machine, 3 October 2015. Accessed May 10, 2019.

30. Regulation (eu) 2016/679 of the european parliament and of the council. https://eur-lex.europa.eu/legal-content/EN/TXT/?uri=CELEX%3A32016R0679, 2016. Accessed June 30, 2019.

31. An analysis of trends in cost of remittance services of remittance prices worldwide. http://remittanceprices.worldbank.org, 2017. Accessed May 10, 2019.

32. Remittance prices worldwide. http://remittanceprices.worldbank.org, 2017. Accessed April 10, 2019.

33. Sec issues investigative report concluding dao tokens, a digital asset, were securities. https://www.sec.gov/corpfin/framework-investment-contract-analysis-digital-assets, 2017. Accessed June 2, 2019.

34. Smart contracts and distributed ledger? A legal perspective. https://www.linklaters.com/en/about-us/news-and-deals/news/2017/smart-contracts-and-distributed-ledger--a-legal-perspective, 2017. Accessed April 5, 2019.

35. https://mdia.gov.mt/wp-content/uploads/2018/10/MDIA.pdf, 2018. Accessed May 16, 2019.

36. https://mdia.gov.mt/wp-content/uploads/2018/10/ITAS.pdf, 2018. Accessed May 16, 2019.

37. Blockchain and the gdpr. Technical report, 2018. Accessed June 17, 2019.

38. Breaking blockchain open, Deloitte's 2018 global blockchain survey. https://www2.deloitte.com/uk/en/pages/innovation/articles/global-blockchain-survey-2018.html, 2018. Accessed April 10, 2019.

39. Crypto-assets report to the g20 on work by the fsb and standard-setting bodies. Report, Financial Stability Board, 2018.

40. Crypto money-laundering. https://www.economist.com/finance-and-economics/2018/04/26/crypto-money-laundering, April 2018. Accessed July 10, 2019.

41. Cryptoassets. https://www.fca.org.uk/consumers/cryptoassets, 2018. Accessed July 12, 2019.

42. Cryptocurrencies: looking beyond the hype. https://www.bis.org/publ/arpdf/ar2018e5.htm, 2018. Accessed June 15, 2019.

43. Evidence submitted by Bank of England (dgc0055). https://data.parliament.uk/writtenevidence/committeeevidence.svc/evidencedocument/treasury-committee/digital-currencies/written/82252.pdf, May 2018. Accessed April 20, 2019.

44. Legal framework for distributed ledger technology and blockchain in switzerland. https://www.mme.ch/fileadmin/files/.../181207_Bericht_Bundesrat_Blockchain_Engl.pdf, 2018. Accessed May 16, 2019.

45. Own initiative report on initial coin offerings and crypto-assets. https://www.esma.europa.eu/, 2018. Accessed July 2, 2019.
46. Project Khokha. https://www.resbank.co.za/Lists/.../8491/SARB_ProjectKhokha%2020180605.pdf, 2018. Accessed May 5, 2019.
47. Smart derivatives contracts: From concept to construction. https://www.kwm.com/en/au/knowledge/insights/smart-derivatives-contracts-from-concept-to-construction-20181004, 2018. Accessed April 5, 2019.
48. Two ico issuers settle sec registration charges, agree to register tokens as securities. https://www.sec.gov/news/press-release/2018-264, 2018. Accessed 2 July 2019.
49. Virtual financial assets act (vfaa). http://www.justiceservices.gov.mt/DownloadDocument.aspx?app=lp&itemid=29079&l=1, 2018. Accessed July 4, 2019.
50. https://www.cnbc.com/2019/04/11/cryptocurrencies-fintech-clearly-shaking-the-system-imfs-lagarde.html, 2019. Accessed June 28, 2019.
51. Blockchain for Zero Hunger. https://innovation.wfp.org/project/building-blocks, 2019. Accessed April 5, 2019.
52. Central banks and distributed ledger technology: How are central banks exploring blockchain today? www3.weforum.org/docs/WEF_Central_Bank_Activity_in_Blockchain_DLT.pdf, 2019. Accessed May 10, 2019.
53. Distributed ledger technical research in Central Bank of Brazil. https://www.bcb.gov.br/htms/public/microcredito/Distributed_ledger_technical_research_in_Central_Bank_of_Brazil.pdf, 2019. Accessed May 10, 2019.
54. Facebook wants to create a global currency, 20 June 2019. Accessed June 17, 2019.
55. Framework for 'investment contract' analysis of digital assets. https://www.sec.gov/corpfin/framework-investment-contract-analysis-digital-assets, 2019. Accessed July 2, 2019.
56. Inclusive deployment of blockchain for supply chains: Part 1? Introduction. www3.weforum.org/docs/WEF_Introduction_to_Blockchain_for_Supply_Chains.pdf, 2019. Accessed May 12, 2019.
57. Nearly a third of millennials say they'd rather own bitcoin than stocks. https://www.bloomberg.com/news/articles/2017-11-08/millennials-ready-to-ditch-stocks-to-keep-bitcoin-rally-alive, 2019. Accessed June 28, 2019.
58. Worldwide Blockchain Spending Forecast to Reach $2.9 Billion in 2019, According to New IDC Spending Guide. https://www.idc.com/getdoc.jsp?containerId=prUS44898819, 2019. Accessed April 10, 2019.
59. Robleh Ali, John Barrdear, Roger Clews, and James Southgate. The economics of digital currencies. *Bank of England Quarterly Bulletin*, 54(3): 276–286, 2014.
60. Michel Rauchs Andrew, Glidden Brian, Gordon, Gina Pieters Martino Recanatini, François Rostand, Kathryn Vagneur, and Bryan Zhang. Distributed ledger technology systems. Technical report, University of Cambridge, Centre for Alternative Finance, 2017.
61. Adam Back. Hashcash a denial of service counter-measure. http://www.hashcash.org/papers/hashcash.pdf, 2002. Accessed May 20, 2019.

62. Elaine Barker. Nist special publication 800-57 recommendation for key management? Part 1: General. https://csrc.nist.gov/publications/detail/sp/800-57-part-1/rev-4/final, 2016. Revision 4.

63. Elaine B. Barker. Sp 800-102. recommendation for digital signature timeliness. Technical report, Gaithersburg, MD, United States, 2009.

64. Christian Barontini and Henry Holden. Proceeding with caution—A survey on central bank digital currency. BIS papers 101, Bank for International Settlements, 2019.

65. Christian Barontini and Henry Holden. Proceeding with caution? A survey on central bank digital currency. *Bank for International Settlements (BIS) Quarterly Review*, January 2019.

66. John Barrdear and Michael Kumhof. The macroeconomics of central bank issued digital currencies. Bank of England working papers 605, Bank of England, 2016.

67. David G.W. Birch. *Before Babylon Beyond Bitcoin*. London Publishing Partnership, London, UK.

68. Aisha Bin Bishr. Dubai: A city powered by blockchain. *Innovations: Technology, Governance, Globalization*, 12(3–4): 4–8, 2019.

69. David Chaum. Blind signatures for untraceable payments. *Advances in Cryptology*, pp. 199–203, 1983.

70. David Chaum. Security without identification: Transaction systems to make big brother obsolete. *Communication of the ACM*, 28(10): 1030–1044, 1985.

71. David Chaum. The dining cryptographers problem: Unconditional sender and recipient untraceability. *Journal of Cryptology*, 1: 65–75, 1988.

72. David Chaum, Amos Fiat, and Naor Moni. Untraceable electronic cash. In *Proceedings of the 8th Annual International Cryptology Conference on Advances in Cryptology*, CRYPTO '88, pp. 319–327. Springer-Verlag, London, UK, 1988.

73. David Chaum, Peter Ryan, and Steve Schneider. A practical, voter-verifiable election scheme, September 2005.

74. David L. Chaum. Untraceable electronic mail, return addresses, and digital pseudonyms. *Communication of the ACM*, 24(2): 84–90, 1981.

75. Christopher D. Clack, Vikram A. Bakshi, and Lee Braine. Smart contract templates: Foundations, design landscape and research directions. *CoRR*, abs/1608.00771, 2016.

76. Marek Dabrowski and Lukasz Janikowski. Virtual currencies and central banks monetary policy: Challenges ahead. https://www.europarl.europa.eu/cmsdata/149900/CASE_FINAL%20publication.pdf, 2018. Accessed May 12, 2019.

77. Akber Datoo. *Legal Data for Banking—Business Optimisation and Regulatory Compliance*. Wiley, Chichester, 2019.

78. S. Murphy and F. Piper. *Cryptography: A Very Short Introduction*. Oxford University Press, Oxford, 2002.

79. Massimo Flore. *How blockchain-based technology is disrupting migrants' remittances: A preliminary assessment*. European Commission, JRC Science for Policy Report: Report, 2018.

80. Sean Foley, Jonathan R. Karlsen, and Tālis J. Putniņš. Sex, drugs, and bitcoin: How much illegal activity is financed through cryptocurrencies? *The Review of Financial Studies*, 32(5): 1798–1853, April 2019.

81. John Kenneth Galbraith. *A Short History of Financial Euphoria*. Penguin, New York, 1994.

82. Jeremy Ginsberg, Matthew H. Mohebbi, Rajan S. Patel, Lynnette Brammer, Mark S. Smolinski, and Larry Brilliant. Detecting influenza epidemics using search engine query data. *Nature*, 457: 1012–1014, 2008.

83. Stuart Haber and W. Scott Stornetta. How to time-stamp a digital document. *Journal of Cryptology*, 3(2): 99–111, January 1991.

84. Garrick Hileman and Michel Rauchs. Global cryptocurrency benchmarking study. Technical report, University of Cambridge, Centre for Alternative Finance, 2017.

85. Jaap-Henk Hoepman. Distributed double spending prevention. *CoRR*, abs/0802.0832, 2008

86. Eric Hughes. The electronic privacy papers chapter A Cypherpunk's Manifesto, pp. 285–287. Wiley, New York, NY, USA, 1997.

87. Melis Jackob. History of encryption. Technical report, Escal Institute of Advanced Technologies (SANS) Technology Institute, 2001. Accessed January 31, 2019.

88. Christoph Jentzsch. Decentralized autonomous organization to automate governance. https://download.slock.it/public/DAO/WhitePaper.pdf, 2016. Accessed May 5, 2019.

89. Fleur Doidge Jessica Exton. Cracking the code on cryptocurrency. https://think.ing.com/reports/cracking-the-code-on-cryptocurrency/, 2018. Accessed March 20, 2019.

90. A. Kerckhoffs. La cryptographie militaire. *Journal des sciences militaires*, IX: 5–38, 1883.

91. Christina Lagarde. A regulatory approach to fintech, 2018.

92. K.M. Martin. *Everyday Cryptography: Fundamental Principles and Applications*. Oxford University Press, Oxford, 2017.

93. Timothy C. May. The crypto anarchist manifesto. https://www.activism.net/cypherpunk/crypto-anarchy.html, 1988. Accessed April 1, 2019.

94. Satoshi Nakamoto. Bitcoin: A peer-to-peer electronic cash system. https://bitcoin.org/en/bitcoin-paper, 2008.

95. Arvind Narayanan, Joseph Bonneau, Edward Felten, Andrew Miller, and Steven Goldfeder. *Bitcoin and Cryptocurrency Technologies: A Comprehensive Introduction*. Princeton University Press, Princeton, NJ, USA, 2016.

96. Arvind Narayanan and Jeremy Clark. Bitcoin's academic pedigree. *ACM Queue*, 15(4): 20, August 2017.

97. O'Dair, Marcus. *Distributed Creativity*. Palgrave Macmillan, Cham, 2019.

98. Florent Pépin and Maria G. Vigliotti. Risk Assessment of the 3DES in ERTMS. pp. 79–92, June 2016.

99. Michel Rauchs, Apolline Blandin, Kristina Klein, Gina Pieters, Martino Recanatini, and Bryan Zhang. 2nd global cryptoasset benchmarking study. Technical report, University of Cambridge, Centre for Alternative Finance, 2018.

100. R.L. Rivest, A. Shamir, and L. Adleman. A method for obtaining digital signatures and public-key cryptosystems. *Communication of the ACM*, 21(2): 120–126, 1978.

101. Martin Roetteler, Michael Naehrig, Krysta M. Svore, and Kristin Lauter. Quantum Resource Estimates for Computing Elliptic Curve Discrete Logarithms. Technical report, 2017.

102. Bruce Schneier. *Applied Cryptography*. Wiley, New York, 1996.

103. Hernando De Soto. *The Mystery of Capital: Why Capitalism Triumphs in the West and Fails Everywhere Else*. Black Swan Book, London, UK, 2001.

104. Nick Szabo. Smart contracts. http://www.fon.hum.uva.nl/rob/Courses/InformationInSpeech/CDROM/Literature/LOTwinterschool2006/szabo.best.vwh.net/smart.contracts.html, 1994. Accessed December 20, 2018.

105. Nick Szabo. Smart contracts: Building blocks for digital markets. http://www.alamut.com/subj/economics/nick_szabo/smartContracts.html, 1996. Accessed December 20, 2018.

106. Paolo Tasca, Maria G. Vigliotti, and Hugo Guang. Risks and challenges of initial coin offerings. *Journal of Digital Banking*, 3(1): 80–94, 2018.

107. Petros Wallden and Elham Kashefi. Cyber security in the quantum era. *Communication of the ACM*, 62(4): 120–120, 2019.

108. D. Yaga, P. Roby Mell, and K. Scarfone. Blockchain technology overview (nistir 8202). Report, National Institute of Standards and Technology (NIST), 2018.

Index

Printed by Printforce, the Netherlands